PERGAMON INTERNATIONAL LIBRARY
of Science, Technology, Engineering and Social Studies

*The 1000-volume original paperback library in aid of education,
industrial training and the enjoyment of leisure*

Publisher: Robert Maxwell, M.C.

SI UNITS IN ENGINEERING
AND TECHNOLOGY

THE PERGAMON TEXTBOOK
INSPECTION COPY SERVICE

An inspection copy of any book published in the Pergamon International Library
will gladly be sent to academic staff without obligation for their consideration for
course adoption or recommendation. Copies may be retained for a period of 60 days
from receipt and returned if not suitable. When a particular title is adopted or
recommended for adoption for class use and the recommendation results in a sale
of 12 or more copies the inspection copy may be retained with our compliments.
The Publishers will be pleased to receive suggestions for revised editions and new
titles to be published in this important international Library.

Other Titles of Interest
in the Pergamon International Library

SI UNITS IN ENGINEERING AND TECHNOLOGY

by

S. H. QASIM

Professor of Mechanical Engineering
Basrah University

PERGAMON PRESS

OXFORD · NEW YORK · TORONTO
SYDNEY · PARIS · FRANKFURT

U.K.	Pergamon Press Ltd., Headington Hill Hall, Oxford OX3 0BW, England
U.S.A.	Pergamon Press Inc., Maxwell House, Fairview Park, Elmsford, New York 10523, U.S.A.
CANADA	Pergamon of Canada Ltd., 75 The East Mall, Toronto, Ontario, Canada
AUSTRALIA	Pergamon Press (Aust.) Pty. Ltd., 19a Boundary Street, Rushcutters Bay, N.S.W. 2011, Australia
FRANCE	Pergamon Press SARL, 24 rue des Ecoles, 75240 Paris, Cedex 05, France
WEST GERMANY	Pergamon Press GmbH, 6242 Kronberg-Taunus, Pferdstrasse 1, Frankfurt-am-Main, West Germany

First edition 1977

Library of Congress Cataloging in Publication Data

Qasim, S.H.
SI units in engineering and technology.

1. International system of units. I. Title.
T50.5.Q37 1977   389'.152   76-27670
ISBN 0-08-021278-6

Printed in Great Britain by A. Wheaton & Co., Exeter

Contents

Preface

More and more countries are adopting the International System of Units—Système International d'Unités or commonly known as SI. It is expected to become the only system for weights and measures throughout the world. This booklet is intended to serve the needs of Education and Industry for better understanding of the SI units. It is hoped that the book will serve as a quick reference for the purpose of conversion of Imperial units to the SI units.

I have great pleasure in expressing my gratitude to the International Standardization Organization (ISO) for their kind permission to reproduce parts of the relevant standards and their ready assistance in any matter concerning the use of SI units. The full standards can be obtained from the standard organization of each country.

Basrah, Iraq   S. H. QASIM

CHAPTER 1

International System of Units (SI description)

1.1 Basic SI units

According to the latest information from International Organization for Standardization (ISO) there are seven basic units from which a wide range of quantities can be derived in the form of products and/or quotient of these basic units. The seven basic units are given below:

No.	Quantity	Name of unit	Recommended unit symbol
1.	Length	metre	m
2.	Mass	kilogram	kg
3.	Time	second	s
4.	Electric current	*ampere	A
5.	Temperature	*kelvin	K
6.	Luminous intensity	candela	cd
7.	Amount of substance	mole	mol

*According to ISO if the unit is named after a person, it should be written in small letters and a capital letter be used for the symbol.

According to ISO (Pub. ISO R.31, 1965) the definition of these basic units is as under:

Metre (m). The metre is the length equal to 1 650 763.73 wavelengths in vacuum of the radiation corresponding to the transition between levels $2p_{10}$ and $5d_5$ of the krypton-86 atom.

Kilogram (kg). The kilogram is the unit of mass: it is equal to the mass of the international prototype of the kilogram.

Second (s). The second is the duration of 9 192 631 770 periods of the radiation corresponding to the transition between the two hyperfine levels of the ground state of the caesium-133 atom.

Ampere (A). The ampere is that constant current which, if maintained in two straight parallel conductors of infinite length, of negligible circular cross-section, and placed one metre apart in vacuum, would produce between these conductors a force equal to 2×10^{-7} newton per metre of length.

Kelvin (K). The kelvin, unit of thermodynamic temperature, is the fraction $\frac{1}{273 \cdot 15}$ of the thermodynamic temperature of the triple point of water.

Candela (cd). The candela is the luminous intensity in the perpendicular direction, of a surface of $\frac{1}{600\,000}$ square metre of a black body at the temperature of freezing platinum under a pressure of 101 325 newtons per square metre.

Mole (mol). The mole is defined as an amount of substance of a system which contains as many elementary units as there are carbon atoms in 0.012 kg (exactly) of the pure nuclide carbon-12.

The supplementary base units radian and steradian are defined also:

Plane angle (radian). The angle subtended at the centre of a circle of radius 1 metre by an arc of length 1 metre along the circumference.

Solid angle (steradian). The solid angle subtended at the centre of a sphere of radius 1 metre by an area of 1 square metre on the surface.

1.2 Magnitude of SI units

The SI system is a decimal system in which calculations using the numeral 10, multiplied or divided by itself is often used. It will be of interest to remember the following simple mathematical rules for such calculations:

$$10^m = 10 \times 10 \times 10 \ldots \text{to } m \text{ factors.}$$

$$10^m \times 10^n = 10^{(m+n)} ,$$

$$10^m \times 10^n \times 10^p = 10^{(m+n+p)} ,$$

$$\frac{10^m}{10^n} = 10^{(m-n)} \qquad \text{if } m > n ,$$

$$= 10^{(n-m)} \qquad \text{if } n > m ,$$

$$(10^m)^n = 10^{(mn)} ,$$

$$10^0 = 1 ,$$

$$10^{-n} = \frac{1}{10^n} ,$$

$$(10a)^m = 10^m \, a^m ,$$

$$\left(\frac{10^x}{10^y}\right)^m = \frac{10^{xm}}{10^{ym}} ,$$

$$(10)^{m/n} = \sqrt[n]{10^m} .$$

To obtain multiples and sub-multiples of the units, standard prefixes have been recommended by the ISO, shown in Table 1.1.

TABLE 1.1. Standard Multiples and Sub-multiples

Multiplication factor		Prefix	Symbol
One million million (billion)	$1\ 000\ 000\ 000\ 000 = 10^{12}$	tera	T
One thousand million	$1\ 000\ 000\ 000 = 10^{9}$	giga	G
One million	$1\ 000\ 000 = 10^{6}$	mega	M
One thousand	$1\ 000 = 10^{3}$	kilo	k
One hundred	$100 = 10^{2}$	hecto	h*
Ten	$10 = 10^{1}$	deca	da*
Unity	$1 = 10^{0}$	–	–
One tenth	$0.1 = 10^{-1}$	deci	d*
One hundredth	$0.01 = 10^{-2}$	centi	c*
One thousandth	$0.001 = 10^{-3}$	milli	m
One millionth	$0.000\ 001 = 10^{-6}$	micro	μ
One thousand millionth	$0.000\ 000\ 001 = 10^{-9}$	nano	n
One million millionth	$0.000\ 000\ 000\ 001 = 10^{-12}$	pico	p
One thousand million millionth	$0.000\ 000\ 000\ 000\ 001 = 10^{-15}$	femto	f
One million million millionth	$0.000\ 000\ 000\ 000\ 000\ 001 = 10^{-18}$	atto	a

*It is suggested that all SI units be expressed in "preferred standard form" in which the multiplier is 10^{3n} where n is a positive or negative whole number. Consequently the use of hecto, deca, deci and centi is to be avoided wherever possible.

1.3 Rules for multiples and sub-multiples

The following rules have been recommended for use of SI units:

1. The basic SI units are to be preferred but it is impracticable to limit the usage to these alone (e.g. it is rather awkward to measure road length and breadth of human hair in metres), therefore, their decimal multiples and sub-multiples are also required.

2. In order to avoid errors in calculations, it is preferable to use coherent units. It is, therefore, strongly recommended that in calculations only SI units are used and not their multiples and sub-multiples. (*Example*: Prefer N/m² × 10^6 to MN/m² or N/mm².)

3. The use of prefixes representing 10 raised to the power "$3n$" is especially recommended where n is a positive or negative integer. (*Example*: μF, mm, kg are preferred and hm, cm and dag to be avoided.)

4. When expressing a quantity by a numerical value of a unit, it is advisable to use quantities resulting in numerical values between 0 and 1000. (*Example*: 15 kN or 15 × 10^3 N and not 15 000 N; 2.75 mm or 2.75 × 10^{-3} m and not 0.002 75 m.)

5. Compound prefixes are not used. [*Example*: write nm (nanometre) and not mμm (millimicrometre).]

6. Multiplying prefixes are printed immediately adjacent to the SI unit symbol with which they are associated. The multiplication of symbols is usually indicated by leaving a small gap between them. (*Example*: mN means millinewton whereas m N would indicate metre newton.)

7. The numbers should be grouped in three on either side of the decimal marker and placing of comma should be avoided. (*Example*: 8 475.2 and not 8,475.2; 0.002 52 and not 0.00252.)

8. A prefix applied to a unit becomes part of that unit and is subject to any applied power. [*Example:* 1 mm³ = 1 (mm)³ = 10^{-9} m³ and not 1 mm³ = 1 m(m)³ = 10^{-3} m³.]

TABLE 1.2. Quantities and Units of Mechanics (R.31 ISO)

Symbol	Quantity
m	Mass
e, ρ	Density (mass density)
d	Relative density
v	Specific volume
p	Momentum
b, p_0, p_θ	Moment of momentum (angular momentum)
I, J	Moment of inertia (dynamic moment of inertia)
F	Force
$G(P, W)$	Weight
γ	Specific weight (weight density)
M	Moment of force; Bending moment
T	Torque, moment of a couple
p	Pressure
σ	Normal stress
τ	Shear stress
e, ϵ	Linear strain (relative elongation)
γ	Shear strain (shear angle)
Θ, θ	Volume strain (bulk strain)
μ, ν	Poisson's ratio, Poisson's number
E	Young's modulus (modulus of elasticity)
G	Shear modulus (modulus of rigidity)
K	Bulk modulus (modulus of compression)
x, κ	Compressibility (bulk compressibility)
I, I_a	Second moment of area (second axial moment of area)
I_p, J	Second polar moment of area
$Z, W, \left(\dfrac{1}{v}\right)$	Section modulus
$\mu(f)$	Coefficient of friction (factor of friction)
$\eta(\mu)$	Viscosity (dynamic viscosity)
γ	Kinematic viscosity
$\sigma, (\gamma)$	Surface tension
A, W	Work
E, W	Energy
E_p, U, V, Φ	Potential energy
E_k, K, T	Kinetic energy
p	Power

TABLE 1.3. Quantities and Units of Periodic and
Related Phenomena

Symbol	Quantity
T	Periodic time
$\tau(T)$	Time constant
f, ν	Exponentially varying quantity; frequency
η	Rotational frequency
ω	Angular frequency
l	Wavelength
$\sigma(\tilde{\nu})$	Wave number
k	Circular wave number
$\log_e(A_1/A_2)$	Natural logarithm of the ratio of two amplitudes
$10 \log_{10}(P_1/P_2)$	Ten times the common logarithm of ratio of two powers
δ	Damping coefficient
Λ	Logarithmic decrement
α	Attenuation coefficient
β	Phase coefficient
γ	Propagation coefficient

TABLE 1.4. Mathematical Signs and Symbols for use in Technology

Sign, Symbol	Quantity
$=$	Equal to
$\neq$	Not equal to
$\equiv$	Identically equal to
$\triangle$	Corresponds to
$\approx$	Approximately equal to
$\rightarrow$	Approaches
$\simeq$	Asymptotically equal to
$\sim$	Proportional to
∞	Infinity
$<$	Smaller than
$>$	Larger than
$\leq$ $\leqslant$ $\leqq$	Less than or equal to
$\geq$ $\geqslant$ $\geqq$	Greater than or equal to
$\ll$	Much smaller than
$\gg$	Much larger than
$+$	Plus
$-$	Minus
$\times$	Multiplied by
$\dfrac{a}{b}$, a/b	a divided by b
$|a|$	Magnitude of a
a^n	a raised to the power n
$a^{\frac{1}{2}} a^{1/2} \sqrt{a}$	square root of a
$a^{\frac{1}{n}} \sqrt[n]{a} \sqrt[n]{a}$	nth root of a
$\bar{a}$ $(<a)$	Mean value of a
$p!$	Factorial p, $1 \times 2 \times 3 \times \ldots \times p$
$\binom{n}{p}$	Binomial coefficient, $\dfrac{n(n-1)\ldots(n-p+1)}{1\times 2\times 3\ldots\times p}$
Σ	Sum
Π	Product
$f(x)$	Function of the variable x
$[f(x)]^b_a$ $f(x)/^b_a$	$f(b) - f(a)$
$\lim_{x\to a} f(x)$; $\lim_{x\to a} f(x)$	The limit to which $f(x)$ tends as x approaches a
Δx	Delta x = finite increment of x
δx	Delta x = variation of x
$\dfrac{df}{dx}$; df/dx; $f'(x)$	Differential coefficient of $f(x)$ with respect to x
$\dfrac{d^n f}{dx^n}$; $f^{(n)}(x)$	Differential coefficient of order n of $f(x)$
$\dfrac{\partial f(x,y,\ldots)}{\partial x}$; $\left(\dfrac{df}{dx}\right)_{y\ldots}$	Partial differential coefficient of $f(x,y,\ldots)$ with respect to x, when $y, \ldots$ are held constant
df	Total differential of f
$\int f(x)\,dx$	Indefinite integral of $f(x)$ with respect to x
$\int_a^b f(x)\,dx$; $\int_a^b f(x)\,dx$	Definite integral of $f(x)$ from $x=a$ to $x=b$

TABLE 1.4. Mathematical Signs and Symbols for use in Technology

Sign, Symbol	Quantity
e	Base of natural logarithms
e^x; exp x	e raised to the power x
$\log_a x$	Logarithm to the base a of x
ln x; $\log_e x$	Natural logarithm (Napierian logarithm) of x
lg^x; log^x; $\log_{10}^x$	Common logarithm of x
sin x	Sine of x
cos x	Cosine of x
tan x; tg x	Tangent of x
cot x; ctg x	Cotangent of x
sec x	Secant of x
cosec x	Cosecant of x
arcsin x	Arc sine of x
arccos x	Arc cosine of x
arctan x; arctg x	Arc tangent of x
arccot x; arcctg x	Arc cotangent of x
arcsec x	Arc secant of x
arccosec x	Arc cosecant of x
sinh x	Hyperbolic sine of x
cosh x	Hyperbolic cosine of x
tanh x	Hyperbolic tangent of x
coth x	Hyperbolic cotangent of x
sech x	Hyperbolic secant of x
cosech x	Hyperbolic cosecant of x
arsinh x	Inverse hyperbolic sine of x
arcosh x	Inverse hyperbolic cosine of x
artanh x	Inverse hyperbolic tangent of x
arcoth x	Inverse hyperbolic cotangent of x
arsech x	Inverse hyperbolic secant of x
arcosech x	Inverse hyperbolic cosecant of x
i, j	Imaginary unity, $\sqrt{-1}$
Re z	Real part of z
Im z	Imaginary part of z
$\lvert z \rvert$	Modulus of z
arg z	Argument of z
z^*	Conjugate of x, complex conjugate of z
$\tilde{A}$	Transpose of matrix A
A^*	Complex conjugate matrix of matrix A
$A^\dagger$	Hermitian conjugate matrix of matrix A
Aa	Vector A
$\lvert A \rvert$, A	Magnitude of vector
A.B	Scalar product
A $\times$ **B**; **A** $\triangle$ **B**	Vector product
∇	Differential vector operator
$\nabla\phi$, grad ϕ	Gradient of ϕ
$\nabla.A$, div A	Divergence of A
$\nabla \times A$, $\nabla\cdot\wedge A$ curl A, rot A $\}$	Curl of A
$\nabla^2\phi$, $\Delta\phi$	Laplacian of ϕ

CHAPTER 2

Derivation of Important Engineering Units

2.1 Force

Mass is the quantity of matter in a body. In the SI, the unit of mass is the kilogram (kg). According to Sir Isaac Newton, the force is proportional to the rate of change of momentum. Expressed mathematically:

$$F \propto \frac{mv - mu}{t},$$

$$\propto m\left(\frac{v - u}{t}\right),$$

$$\propto m \times a$$

or
$$F = k \times m \times a \tag{i}$$

where k is the proportionality constant.

The SI system of units is rationalized and coherent. It is rationalized because for any one physical quantity, only one measurement unit is essential and its entire structure is derived from no more than seven arbitrarily defined basic units. The system of units is "coherent" if the product or quotient of any two unit quantities in the system is the unit of the resultant quantity. Hence the unit of derived quantity in equation (i) is obtained from unit values, i.e. the proportionality constant k should be unity. Consequently we get

$$F = m \times a$$

or 1 unit of force $=$ 1 unit of mass (1 kg) $\times$ 1 unit of acceleration (1 m/s^2),
$$= 1 \text{ kg m/s}^2.$$

The unit of force is called "newton" to accredit the work of Sir Isaac Newton as the pioneer. It is a standard practice in SI units to write small letters for name of a person in full and a capital letter for the symbol. Hence:

$$1 \text{ newton} = 1 \text{ N} = 1 \text{ kg m/s}^2.$$

7

A newton is then defined as the force which when acting on a mass of 1 kilogram gives to that mass an acceleration of 1 metre per second per second.

The value of gravitational acceleration varies from one place to another but it has been agreed to accept an average at 9.806 65 m/s² (for practical purposes 9.81 m/s²). So if a mass of 1 kilogram is allowed to fall freely, the force experienced by it in the SI system of units would be:

$$F = m \times a$$

$$= m \times g \; (g = \text{acceleration due to gravity})$$

$$\text{let } g = 9.81 \text{ m/s}^2 \text{ approximately}$$

$$\text{or } F = 1 \text{ kg} \times 9.81 \text{ m/s}^2$$

$$= 9.81 \text{ kg m/s}^2$$

$$1 \text{ kgf} = 9.81 \text{ N (newtons)}.$$

In other words a mass of 1/8.91 kg = 102 grams falling freely under gravity ($g = 9.81$ m/s²) will experience a force of 1 newton.

2.2. Stress and pressure

Stress or pressure is defined as the force per unit area. In the SI system it would be expressed mathematically as

$$\sigma \text{ or } p = \frac{F}{A}$$

$$= \frac{N}{m^2} = N/m^2.$$

The unit for pressure or stress is the pascal and is same as newtons per square metre. It may also be mentioned that for most of the engineering materials, their design stresses are in the range of 10^6 N/m² or MN/m² and this can be readily converted to N/mm² as $1 \text{ MN/m}^2 = 1 \text{ N/mm}^2$.

2.3. Work and energy

The SI is a coherent system of units, therefore, from the basic principles:

$$\text{Work done} = \text{force} \times \text{distance moved}$$

$$= \text{newton} \times \text{metres}$$

$$= N \text{ m}.$$

It is preferable to write newton-metres as N m and not mN. In the latter case the confusion may arise with the millinewton.

$$\text{One newton metre } = \ 1 \, \text{N m}$$

$$= \ 1(\text{kg m/s}^2) \times (\text{m})$$

$$= \ 1 \ \text{kg m}^2/\text{s}^2$$

$$= \ 1 \, \text{J}.$$

Thus 1 joule of work is done when point of application of a force of 1 newton is displaced through a distance of 1 metre in the direction of that force.

Similarly for energy the unit is the joule:

$$\text{kinetic energy } = \ \tfrac{1}{2} \, mV^2,$$

$$\text{potential energy } = \ mgh \, ,$$

$$\text{units of energy } = \ \text{kg m}^2/\text{s}^2 \ = \ \text{J (joule).}$$

2.4. Power

Power is defined as the rate of doing work, i.e. work done per unit time. Hence:

$$P \ = \ \frac{\text{work done}}{\text{time}} \ = \ \text{N m/s}$$

$$= \ \text{J/s}$$

$$= \ \text{W}$$

$$= \ \text{watt (kg m}^2/\text{s}^3).$$

In common engineering usage, watt is rather a small unit, hence kilowatt (kW) equal to 1000 watts is usually used.

The three basic units of importance to the mechanical engineer are newton, joule and watt (N, J and W). The apparent advantage achieved by SI units is that now the unit of work and energy, joule and watt, respectively, would be same throughout mechanics, physics, mechanical and electrical engineering. Consequently all the branches of science would share the same coherent unit of energy.

The unit of horsepower (h.p.) would now become obsolete and will be replaced by watt. The horsepower was a historic unit dating back to the times when the usual external form of power available to man was a horse. It was based on the assumption that a horse could travel 2.5 miles per hour for 8 hours a day pulling a load of 150 lb vertically out of a shaft by means of rope. This is equivalent to work done at the rate of 33 000 ft-lb per minute or 550 ft-lb per second. The SI unit of power watt (W) is related to horsepower thus:

$$1 \text{ h.p. } = \ 746 \, \text{W}.$$

2.5. Temperature

The SI unit for temperature is the kelvin (K) scale in which the triple point of water is supposed to have a temperature 273.15 K. The degree celsius or degree centigrade (°C) in which the temperature of ice is assumed zero would, however, continue to be used for everyday application. The unit degree celsius is exactly equal to the unit kelvin.

A formal definition of the fahrenheit scale does not exist but it is generally recognized that the temperature of melting ice is 32°F and water boils under standard atmosphere at 212°F. The fahrenheit scale of temperature would soon become obsolete. However, the temperature expressed in degrees fahrenheit can be readily converted to degrees celsius and kelvin by the following relationships:

$$\text{degrees F} = \frac{9 \times \text{degrees C}}{5} + 32,$$

$$0°C = 273.15 \text{ K}.$$

CHAPTER 3

Derived SI Units in Science and Engineering

TABLE 3.1. Applied mechanics, mechanical engineering

Quantity	SI unit	Symbol	Definition
Force	newton	N	kg m/s^2
Work, energy, quantity of heat	joule	J	N m
Power, heat flow rate	watt	W	J/s
Moment of force	newton metre	–	N m
Pressure, stress	pascal	Pa	N/m^2
Temperature (basic)	kelvin	K	–
Temperature (common use)	celsius	°C	0°C = 273.15 K
Surface tension	newtons per metre	–	N/m
Thermal coefficient of linear expansion	reciprocal kelvin	–	K^{-1}
Heat flux density, irradiance	watt per square metre	–	W/m^2
Thermal conductivity	watt per metre kelvin	–	W/m K
Coefficient of heat transfer	watt per square metre kelvin	–	W/m^2 K
Heat capacity	joules per kelvin	–	J/K
Specific heat capacity	joules per kilogram kelvin	–	J/kg K
Entropy	joules per kelvin	–	J/K
Specific entropy	joules per kilogram kelvin	–	J/kg K
Specific energy: specific latent heat	joules per kilogram	–	J/kg
Viscosity (kinematic)	metre squared per second	–	m^2/s
Viscosity (dynamic)	pascal second	–	Pa s

TABLE 3.2. Electrical units

Quantity	SI unit	Symbol	Definition
Electric resistance	ohm	Ω	V/A
Electric charge	coulomb	C	A s
Electric potential difference or voltage or e.m.f.	volt	V	W/A
Electric conductance	siemens	S	A/V
Electric capacitance	farad	F	A s/V
Luminance	candela per square metre (nit)	–	cd/m^2
Illumination	lux	lx	lm/m^2
Luminous flux	lumen	lm	cd sr
Frequency	hertz	Hz	s^{-1}
Electric field strength	volts per metre	–	V/m
Electric flux density	coulombs per square metre	–	C/m^2

TABLE 3.3. Magnetic units

Quantity	SI unit	Symbol	Definition
Magnetic flux	weber	Wb	V s
Inductance	henry	H	V s/A
Magnetic field strength	amperes per metre	—	A/m
Intensity of magnetization	amperes per metre	—	A/m
Magnetic flux density	tesla	T	Wb/m^2

CHAPTER 4

SI Units Conversion Tables
(Arranged according to subject categories)

TABLE 4.1. SI Units (Conversion Factors)
Note: The underlined conversion factors are exact in value

Length

1 angstrom (Å)	$= \underline{10^{-10}\,\text{m}}$
1 microinch (1 μin)	$= \underline{0.025\ 4\ \mu\text{m}}$
1 thou; milli-inch or mil	$= \underline{25.4\ \mu\text{m}}$
1 inch (in)	$= \underline{25.4\ \text{mm}}$
1 foot (ft)	$= \underline{0.304\ 8\ \text{m}}$
1 yard (yd)	$= \underline{0.914\ 4\ \text{m}}$
1 fathom (6 ft)	$= \underline{1.828\ 8\ \text{m}}$
1 furlong (220 yd)	$= \underline{0.201\ 168\ \text{km}}$
1 mile	$= \underline{1.609\ 344\ \text{km}}$

Area

1 circular mil $\left(\dfrac{\pi}{4} \times 10^{-6}\,\text{in}^2 \right)$	$= 506.707\ \mu\text{m}^2$
1 square inch (in^2)	$= \underline{645.16\ \text{mm}^2}$
1 square foot (ft^2)	$= 0.092\ 903\ \text{m}^2$
1 square yard (yd^2)	$= 0.836\ 127\ \text{m}^2$
1 acre (4840 yd^2)	$= 4\ 046.86\ \text{m}^2$
1 square mile (mile^2)	$= 2.589\ 99\ \text{km}^2$

Capacity and volume

1 cubic inch (in^3)	$= 16.387\ 1\ \text{cm}^3$
1 cubic foot (ft^3)	$= 0.028\ 3168\ \text{m}^3$
1 cubic yard (yd^3)	$= 0.764\ 555\ \text{m}^3$
1 fl oz (1/20 pint)	$= 28.413\ \text{cm}^3$
1 gill (¼ pint)	$= 0.142\ 065\ \text{dm}^3$
1 pint (pt)	$= 0.568\ 261\ \text{dm}^3$
1 quart (qt)	$= 1.136\ 52\ \text{dm}^3$
1 imperial gallon	$= 4.546\ 09\ \text{dm}^3$
1 U.S. gallon (231 in^3)	$= 3.785\ 41\ \text{dm}^3$

Second moment of area

1 ft^4	$= 0.008\ 630\ 97\ \text{m}^4$
1 in^4	$= 0.416\ 231 \times 10^{-6}\ \text{m}^4$
1 in^3 (section modulus)	$= 16.387\ 1 \times 10^{-6}\ \text{m}^3$

TABLE 4.1. (continued).

Velocity and acceleration

1 in/min	= 0.423 333 mm/s
1 in/s	= 25.4 mm/s
1 ft/min	= 0.005 08 m/s
1 ft/s	= 0.3048 m/s
1 mile/h (m.p.h.)	= 1.609 34 km/h
1 mile/h (m.p.h.)	= 0.447 040 m/s
1 ft/s^2	= 0.3048 m/s^2
1 in/s^2	= 25.4 m/s^2

Mass

1 grain (gr)	= 64.7989 mg
1 dram (dr)	= 1.771 85 g
1 ounce (oz)	= 28.3495 g
1 pound (lb)	= 0.453 592 427 7 kg
1 stone (14 lb)	= 6.350 29 kg
1 quarter (qr) (28 lb)	= 12.7006 kg
1 slug	= 14.5939 kg
1 cental (ctl) (100 lb)	= 45.3592 kg
1 hundredweight (cwt) (112 lb)	= 50.8023 kg
1 ton (2240 lb)	= 1016.05 kg
1 U.S. ton (2000 lb)	= 907.18 kg

Mass per unit length

1 oz/in.	= 1.116 12 kg/m
1 lb/in.	= 17.8580 kg/m
1 lb/ft	= 1.488 16 kg/m
1 lb/yd	= 0.496 055 kg/m
1 ton/1000 yd	= 1.111 16 kg/m
1 ton/mile	= 0.631 342 t/km (Mg/km)

Mass per unit area

1 oz/ft^2	= 305.152 g/m^2
1 oz/yd^2	= 33.9057 g/m^2
1 lb/in^2	= 703.070 kg/m^2
1 lb/ft^2	= 4.882 43 kg/m^2

Specific volume

1 ft^3/ton	= 0.027 8696 dm^3/kg
1 ft^3/lb	= 62.428 dm^3/kg
1 in^3/lb	= 36.1273 cm^3/kg
1 gal/lb	= 10.0224 dm^3/kg

Mass rate of flow

1 ton/h	= 1.016 05 t/h (Mg/h)
1 lb/h	= 0.453 592 kg/h
1 lb/h	= 0.125 998 g/s
1 lb/min	= 7.559 87 g/s
1 lb/s	= 0.453 592 kg/s

Volume rate of flow

1 ft^3/s (cusec)	= 0.028 3168 m^3/s
1 gal/h	= 4.546 09 dm^3/h
1 gal/min	= 0.075 7682 dm^3/s
1 gal/s	= 4.546 09 dm^3/s

TABLE 4.1. (continued).

___

Fuel consumption

1 gal/mile	= 2.824 81 dm³/km
1 U.S. gal/mile	= 2.352 15 dm³/km
1 mile/gal	= 0.354 006 km/dm³
1 mile/U.S. gal	= 0.425 144 km/dm³

Density

1 lb/in³	= 27.6799 g/cm³
1 lb/in³	= 27.6799 Mg/m³
1 lb/ft³	= 16.0185 kg/m³
1 slug/ft³	= 515.379 kg/m³
1 lb/gal	= 99.7763 kg/m³
1 ton/yd³	= 1328.94 kg/m³

Force

1 poundal (pdl)	= 0.138 255 N
1 ozf	= 0.278 014 N
1 lbf	= 4.448 22 N
1 tonf	= 9.964 02 kN

Force per unit length

1 lbf/in	= 175.127 N/m
1 lbf/ft	= 14.5939 N/m
1 tonf/ft	= 32.6903 kN/m

Moment of force and torque

1 lbf in	= 0.112 985 N m
1 pdl ft	= 0.042 140 N m
1 lbf ft	= 1.355 82 N m
1 tonf ft	= 3.037 03 kN m

Moment of inertia

1 slug ft²	= 1.355 82 kg m²
1 oz in²	= 18.290 kg mm²
1 lb in²	= 292.640 kg mm²
1 lb ft²	= 0.042 1401 kg m²

Pressure and stress

1 pdl/ft²	= 1.488 16 Pa
1 lbf/in² (psi)	= 6.894 76 kPa
1 lbf/ft²	= 47.8803 Pa
1 tonf/in² (tsi)	= 15.4443 MPa
1 tonf/ft²	= 107.252 kPa
1 ft water	= 2.989 07 kPa
1 in water	= 249.089 Pa
1 in mercury	= 3.386 39 kPa
1 bar	= 10^5 Pa

Dynamic viscosity

1 lbf h/ft²	= 0.172 369 MPa s
1 lbf s/ft²	= 47.8803 Pa s
1 pdl s/ft²	= 1.488 16 Pa s
1 pdl s/ft²	= 1488.16 cP (centipoise)
1 lb/ft s	= 1.488 16 kg/m s
1 lb/ft s	= 1488.16 cP
1 slug/ft s	= 47.8803 km/s

TABLE 4.1. (continued).

Kinematic viscosity

1 ft^2/h	= 0.092 903 m^2/h
1 ft^2/h	= <u>25.8064 cSt</u> (centistokes)
1 ft^2/s	= 0.092 903 m^2/s
1 in^2/h	= 0.179 211 $\times$ 10^{-6} m^2/s
1 in^2/h	= <u>6.4516 cm^2/h</u>
1 in^2/s	= 645.16 mm^2/s
1 in^2/s	= <u>645.16 cSt</u>
1 in^2/s	= 10.7527 mm^2/s
1 in^2/min	= 10.7527 cSt

Energy and work

1 ft pdl	= 0.042 140 J
1 ft lbf	= 1.355 82 J
1 Btu	= 1.055 06 kJ
1 kW h	= <u>3.6 MJ</u>
1 horsepower-hour (h.p.h.)	= 2.684 52 MJ
1 therm	= 105.506 MJ

Power

1 horsepower (hp)	= 745.700 W
1 ft lbf/s	= 1.355 82 W

Heat-flow rate

1 Btu/h	= 0.293 071 W
1 ton of refrigeration (288 000 Btu/day)	= 3.516 85 kW

Thermal conductivity

1 Btu ft/ft^2 h °F	= 1.730 73 W/m °C
1 Btu in/ft^2 h °F	= 0.144 228 W/m °C
1 Btu in/ft^2 s °F	= 519.220 W/m °C

Temperature

1 deg F (fahrenheit)	= $\dfrac{5}{9}$ deg C
T K (kelvin)	= T°C + 273.15
T R (rankine)	= T°F + 459.67
T°C	= (°F − 32) $\times \dfrac{5}{9}$

Illumination

1 lumen/ft^2 (ft-candle)	= 10.7639 lx
1 cd/ft^2	= 10.7639 cd/m^2
1 cd/in^2	= 1550.00 cd/m^2
1 ft lambert	= 3.426 26 cd/m^2

TABLE 4.1. (continued).

Electric Units $\left\{ \begin{array}{l} \text{(CGS units to SI units) "c" is velocity of light in free} \\ \text{space and numerically equal to} \\ 2.997\ 925 \times 10^8 \text{ m/s)} \end{array} \right.$

Current	1 e.m.u.	=	10 A
	1 e.s.u.	=	$1/10c$ A
e.m.f. (potential	1 e.m.u.	=	10^{-8} V
difference)	1 e.su.	=	$10^{-6}\,c$ V
Field strength	1 e.m.u.	=	10^{-6} V/m
	1 e.s.u.	=	$10^{-4}\,c$ V/m
Charge (quantity of electricity)			
	1 e.m.u.	=	10 C
	1 e.s.u.	=	$1/10c$ C
Resistance	1 e.m.u.	=	10^{-9} Ω
	1 e.s.u.	=	$10^{-5}\,c^2$ Ω
Capacitance	1 e.m.u.	=	10^9 F
	1 e.s.u.	=	$10^5/c^2$ F
Inductance	1 e.m.u.	=	10^{-9} H
Magnetic flux	1 e.m.u. (maxwell)	=	10^{-8} Wb
Magnetic flux density	1 e.m.u. (gauss)	=	10^{-4} T
Permeability	1 e.m.u.	=	$4\pi \times 10^{-7}$ H/m
Magnetic field strength	1 e.m.u. (oersted)	=	$10^3/4\pi$ A/m
Electric flux density	1 e.m.u.	=	10^5 C/m²
	1 e.s.u.	=	$10^3/c$ C/m²
Permittivity	1 e.s.u.	=	$10^7/4\pi c^2$ F/m

CHAPTER 5

SI Units Conversion Tables (Arranged alphabetically)

TABLE 5.1. Conversion Factors
Note: The underlined conversion factors are exact in value

Unit	SI equivalent
acre	4046.86 m^2
ampere, A	1 A
ampere/metre, A/m	1 A m^{-1}
angstrom, Å	10^{-10} m
area	100 m^2
assay ton	32.6667 g
atmosphere, standard; atm	101 325 N m^{-2}
bar, b	10^5 N m^{-2}
barn	10^{-28} m^2
biot	10 A
board foot (timber)	0.002 359 74 m^3
Bohr radius	5.291 67 $\times$ 10^{-11} m
Bohr magneton	9.2732 $\times$ 10^{-24} A m^2
British thermal unit, Btu	1.055 06 kJ
British thermal unit/cubic foot, Btu/ft^3	37.2589 kJ m^{-3}
British thermal unit/cubic foot degree F, Btu/ft^3 °F	67.0661 kJ m^{-3} K^{-1}
British thermal unit foot/square foot hour degree F, Btu ft/ft^2 h °F	1.730 73 W m^{-1} K^{-1}
British thermal unit/hour, Btu/h	0.293 071 W
British thermal unit inch/square foot hour degree F, Btu in/ft^2 h °F	0.144 228 W m^{-1} K^{-1}
British thermal unit inch/square foot second degree F, Btu in/ft^2 s °F	519.220 W m^{-1} K^{-1}
British thermal unit/pound, Btu/lb	2326 J kg^{-1}
British thermal unit/pound degree F, Btu/lb °F	4186.8 J kg^{-1} K^{-1}
British thermal unit/pound degree R, Btu/lb °R	4186.8 J kg^{-1} K^{-1}
British thermal unit/square foot hour, Btu/ft^2 h	3.154 59 W m^{-2}
British thermal unit/square foot hour degree F, Btu/ft^2 h °F	5.678 26 W m^{-2} K^{-1}
bushel, bu	0.036 3687 m^3
calorie (international table), cal	4.1868 J
calorie, 15°C, cal$_{15}$	4.1855 J
calorie, thermochemical	4.184 J

TABLE 5.1. (continued).

Unit	SI equivalent
calorie centimetre/square centimetre second degree C, cal cm/cm² s °C	$418.68 \text{ W m}^{-1} \text{ K}^{-1}$
calorie/gram, cal/g	$4.1868 \text{ kJ kg}^{-1}$
calorie/gram degree C, cal/g °C	$4.1868 \text{ J g}^{-1} \text{ K}^{-1}$
calorie/second, cal/s	4.1868 W
calorie/square centimetre second, cal/cm² s	41.868 kW m^{-2}
calorie/square centimetre second degree C, cal/cm² s °C	$41.868 \text{ kW m}^{-2} \text{ K}^{-1}$
candela, cd	1 cd
candela/square foot, cd/ft²	$10.7639 \text{ cd m}^{-2}$
candela/square inch, cd/in²	$1\,550.00 \text{ cd m}^{-2}$
candela/square metre, cd/m²	1 cd m^{-2}
carat, metric	$2 \times 10^{-4} \text{ kg}$
celsius degree, °C	$(T/°C + 273.15) \text{ K}$
cental	45.3592 kg
centigrade degree	$(T/°C + 273.15) \text{ K}$
centigram, cg	10^{-5} kg
centimetre, cm	10^{-2} m
centimetre/second squared, cm/s²	10^{-2} m s^{-2}
centipoise, cP	1 mN s m^{-2}
centistokes, cSt	$10^{-6} \text{ m}^2 \text{ s}^{-1}$
chain	20.1168 m
chain, engineer's	30.48 m
chaldron	1.30927 m^2
cheval vapeur	735.499 W
circular mil	$506.707 \text{ } \mu\text{m}^2$
clusec	$1.33322 \text{ } \mu\text{N ms}^{-1}$
cord (timber)	3.62456 m^3
coulomb, C	1 C
coulomb/square metre, C/m²	1 C m^{-2}
cubic centimetre, cm³	10^{-6} m^3
cubic decimetre, dm³	10^{-3} m^3
cubic foot, ft³	0.0283168 m^3
cubic foot/pound, ft³/lb	$0.0624280 \text{ m}^3 \text{ kg}^{-1}$
cubic foot/second, ft³/s	$0.0283168 \text{ m}^3 \text{ s}^{-1}$
cubic foot/ton, ft³/ton	$0.0278696 \text{ dm}^3 \text{ kg}^{-1}$
cubic inch, in³	16.3871 cm^3
cubic inch/pound, in³/lb	$36.1273 \text{ cm}^3 \text{ kg}^{-1}$
cubic metre, m³	1 m^3
cubic millimetre, mm³	1 mm^3
cubic yard, yd³	0.764555 m^3
curie, Ci	$37 \times 10^9 \text{ s}^{-1}$
cusec	$0.0283168 \text{ m}^3 \text{ s}^{-1}$
cycle/second, c/s	1 Hz
debye, D	$\dfrac{(10^{-19}/c_0) \text{ C m}}{= 3.335640 \times 10^{-30} \text{ Cm}}$
decanewton/square millimetre, daN/mm²	1 daN mm^{-2}
decare	1000 m^2

TABLE 5.1. (continued).

Unit	SI equivalent
decigramme, dg	10^{-4} kg
decilitre, dl	10^{-4} m^3
decimetre, dm	0.1 m
degree (angle), °	$\pi/180$ rad
degree celsius, °C	(T/°C + 273.15) K
degree centigrade	as degree Celsius
degree fahrenheit, °F	5/9 (T/°F + 459.67) K
degree rankine, °R	5/9 (T/°R) K
drachm, fluid	3551.63 mm^3
dram (avoirdupois), dr	1.771 85 g
dyne, dyn	10^{-5} N
dyne/centimetre, dyn/cm	0.001 N m^{-1}
dyne/square centimetre, dyn/cm^2	0.1 N m^{-2}
electromagnetic (e.m.u.) cgs units of the following quantities:	
capacitance, e.m.u. cgs	10^9 F
conductance, e.m.u. cgs	10^9 Ω^{-1}
electric charge, e.m.u. cgs	10 C
electric current, e.m.u. cgs	10 A
electric field strength, e.m.u. cgs	10^{-6} V m^{-1}
electric flux density, e.m.u. cgs	10^5 C m^{-2}
electric potential, e.m.u. cgs	10^{-8} V
electromotive force, e.m.f., e.m.u. cgs	10^{-8} V
inductance, e.m.u. cgs	10^{-9} H
magnetic field strength, e.m.u. cgs (oersted)	$10^3/4\pi$ A m^{-1}
magnetic flux, e.m.u. cgs	10^{-8} Wb
magnetic flux density, e.m.u. cgs (gauss)	10^{-4} T
permeability, e.m.u. cgs	$4\pi \times 10^{-7}$ H m^{-1}
resistance, e.m.u. cgs	10^{-9} Ω
electron volt, eV	1.602 10 $\times$ 10^{-19} J
electrostatic (e.s.u.) cgs units of the following quantities:	
capacitance, e.s.u. cgs	$10^9/c_0^2$ * F
conductance, e.s.u. cgs	$10^9/c_0^2$ Ω^{-1}
electric charge, e.s.u. cgs	$10/c_0$ C
electric current, e.s.u. cgs	$10/c_0$ A
electric field strength, e.s.u. cgs	$10^{-6}c_0$ V m^{-1}
electric flux density, e.s.u. cgs	$10^5/c_0$ C m^{-2}
electric potential, e.s.u. cgs	10^{-8} c_0 V
electromotive force, e.m.f., e.s.u. cgs	10^{-8} c_0 V
permittivity, e.s.u. cgs	$10^{11}/4\pi c_0^2$ F m^{-1}
resistance, e.s.u. cgs	$10^{-9} c_0^2$ Ω
erg	10^{-7} J
fahrenheit degree, °F	*see* degree fahrenheit
farad, F	1 F
farad/metre, F/m	1 F m^{-1}

* c_0 is the true value of velocity of light in free space = 2.977 925 $\times$ 10^8 m/s.

TABLE 5.1. (continued).

Unit	SI equivalent
fathom	1.828 8 m
fluid drachm	3 551.63 mm^3
fluid ounce, fl oz	28.4131 cm^3
foot, ft	0.3048 m
foot-candle, lm/ft^2	10.7639 1x
foot cubed, ft^3	0.028 3168 m^3
foot-lambert	3.426 26 cd m^{-2}
foot/minute, ft/min	0.005 08 m s^{-1}
foot of water (conventional pressure unit)	2989.07 N m^{-2}
foot poundal, ft pdl	0.042 1401 J
foot pound-force, ft lbf	1.355 82 J
foot pound-force/pound, ft lbf/lb	2.989 07 J kg^{-1}
foot pound-force/pound degree F, ft lbf/lb °F	5.380 32 J kg^{-1} K^{-1}
foot pound-force/second, ft lbf/s	1.355 82 W
foot/second, ft/s	0.3048 m s^{-1}
foot/second squared, ft/s^2	0.3048 m s^{-2}
foot squared/hour, ft^2/h	25.8064 × 10^{-6} m^2 s^{-1}
foot squared/second, ft^2/s	0.092 9030 m^2 s^{-1}
foot to the fourth, ft^4	0.008 630 97 m^4
franklin	10/c_0 C
	= 3.335 640 × 10^{-10} C
frigorie	4.1855 kJ h^{-1}
furlong	0.201 168 km
galileo, Gal	0.01 m s^{-2}
gallon, gal	4.546 09 dm^3
gallon/hour, gal/h	4.546 09 dm^3 h^{-1}
gallon/mile, gal/mile	2.824 81 dm^3 km^{-1}
gallon/pound, gal/lb	10.0224 dm^3 kg^{-1}
gallon/second, gal/s	4.546 09 dm^3 s^{-1}
gallon (U.S.), U.S. gal	3.785 41 dm^3
gallon (U.S.)/mile, U.S. gal/mile	2.352 15 dm^3 km^{-1}
gauss, G	10^{-4} T
gigahertz, GHz	1 GHz
gigajoule, GJ	1 GJ
giganewton/square metre, GN/m^2	1 GN m^{-2}
gigawatt, GW	1 GW
gill	0.142 065 dm^3
grade	π/200 rad
grain, gr	64.7989 mg
grain/gallon, gr/gal	14.2538 mg dm^{-3}
grain/100 cubic feet, gr/100 ft^3	0.022 8835 g m^{-3}
gram, g	10^{-3} kg
gram/cubic centimetre, g/cm^3	1000 kg m^{-3}
gram/litre, g/l	1 kg m^{-3}
gram/millilitre, g/ml	1000 kg m^{-3}
hectare, ha	10^4 m^2
hectolitre, hl	0.1 m^3
hectopieze	10^5 Pa

TABLE 5.1. (continued).

Unit	SI equivalent
henry, H	1 H
henry/metre, H/m	1 H m^{-1}
hertz, Hz	1 Hz
horsepower, hp	745.700 W
horsepower, metric	735.499 W
horsepower hour, hp h	2.684 52 MJ
hour, h	3600 s
hundredweight, cwt	50.8023 kg
inch, in.	25.4 mm
inch cubed, in^3	16.3871 cm^3
inch/minute, in/min	$0.423 \text{ } 333 \text{ mm s}^{-1}$
inch of mercury	3.386 39 kPa
inch of water	249.089 Pa
inch/second, in/s	25.4 mm s^{-1}
inch squared/hour, in^2/h	$0.179 \text{ } 211 \times 10^{-6} \text{ m}^{-2} \text{ s}^{-1}$
inch squared/second, in^2/s	$645.16 \text{ mm}^2 \text{ s}^{-1}$
inch to the fourth, in^4	$0.416 \text{ } 231 \times 10^{-6} \text{ m}^4$
international nautical mile, n mile	1.852 km
iron (boot and shoe trade)	0.53 mm
joule, J	1 J
joule/degree celsius, J/°C	1 J K^{-1}
joule/kelvin, J/K	1 J K^{-1}
joule/kilogram, J/kg	1 J kg^{-1}
joule/kilogram degree celsius, J/kg °C	$1 \text{ J kg}^{-1} \text{ K}^{-1}$
joule/kilogram kelvin, J/kg K	$1 \text{ J kg}^{-1} \text{ K}^{-1}$
joule/second, J/s	1 J s^{-1}
kelvin, K	1 K
kilobar, kbar or kb	100 MPa
kilocalorie, kcal	4.1868 kJ
kilocalorie/cubic metre, kcal/m^3	4.1868 kJ m^{-3}
kilocalorie/cubic metre degree C, kcal/m^3 °C	$4.1868 \text{ kJ m}^{-3} \text{K}^{-1}$
kilocalorie/hour, kcal/h	1.163 W
kilocalorie/kilogram, kcal/kg	$4.1868 \text{ kJ kg}^{-1}$
kilocalorie/kilogram degree C, kcal/kg °C	$4.1868 \text{ kJ kg}^{-1} \text{ K}^{-1}$
kilocalorie/kilogram kelvin, kcal/kg K	$4.1868 \text{ kJ kg}^{-1} \text{ K}^{-1}$
kilocalorie metre/square metre hour degree C, kcal m/m² h °C	$1.163 \text{ W m}^{-1} \text{ K}^{-1}$
kilocalorie/square metre hour, kcal/m² h	1.163 W m^{-2}
kilocalorie/square metre hour degree C, kcal/m² h °C	$1.163 \text{ W m}^{-2} \text{ K}^{-1}$
kilogram, kg	1 kg
kilogram/cubic decimetre, kg/dm^3	10^3 kg m^{-3}
kilogram/cubic metre, kg/m^3	1 kg m^{-3}
kilogram/litre, kg/l	1000 kg m^{-3}
kilogram metre squared, kg m²	1 kg m^2

TABLE 5.1. (continued).

Unit	SI equivalent
kilogram metre squared/second, kg m²/s	$1 \text{ kg m}^2 \text{ s}^{-1}$
kilogram metre/second, kg m/s	1 kg m s^{-1}
kilogram-force, kgf	$9.806 \ 65 \text{ N}$
kilogram-force/square centimetre, kgf/cm²	$98.0665 \text{ kN m}^{-2}$
kilogram-force metre, kgf m	$9.806 \ 65 \text{ J}$
kilogram-force/square metre, kgf/m²	$9.806 \ 65 \text{ N m}^{-2}$
kilogram-weight	$9.806 \ 65 \text{ N}$
kilohertz, kHz	1 kHz
kilojoule, kJ	1 kJ
kilojoule/kilogram, kJ/kg	1 kJ kg^{-1}
kilojoule/kilogram degree C, kJ/kg °C	$1 \text{ kJ kg}^{-1} \text{ K}^{-1}$
kilojoule/kilogram kelvin, kJ/kg K	$1 \text{ kJ kg}^{-1} \text{ K}^{-1}$
kilometre, km	1 km
kilometre/hour, km/h	1 km h^{-1}
kilometre/litre, km/l	1 km dm^{-3}
kilonewton, kN	1 kN
kilonewton metre, kN m	1 kN m
kilonewton/square metre, kN/m²	1 kN m^{-2}
kilopond, kp	$9.806 \ 65 \text{ N}$
kilosecond, ks	1 ks
kilovolt/inch, kV/in.	$39.370 \ 1 \text{ kV m}^{-1}$
kilowatt, kW	1 kW
kilowatt hour, kW h	3.6 MJ
kilowatt/square metre, kW/m²	1 kW m^{-2}
kilopound; kip (1000 lb)	$453.592 \ 37 \text{ kg}$
kilopound per square inch (ksi)	6894.76 kPa
knot	$1.853 \ 184 \text{ km h}^{-1}$
knot, international, kn (1 n mile/h)	1.852 km h^{-1}
link	$0.201 \ 168 \text{ m}$
litre, l	$1 \text{ dm}^3 = 10^{-3} \text{ m}^{-3}$
litre atmosphere, l atm	101.328 J
lumen, lm	1 lm
lumen/square foot, lm/ft²	$10.763 \ 9 \text{ lx}$
lumen/square metre, lm/m²	1 lx
lusec	$133.322 \ \mu\text{N m s}^{-1}$
lux, lx	1 lx
maxwell, Mx	10^{-8} Wb
megabar, Mbar or Mb	100 GN m^{-2}
megagram, Mg	1 Mg
megahertz, MHz	1 MHz
megajoule, MJ	1 MJ
megajoule/kilogram, MJ/kg	1 MJ kg^{-1}

TABLE 5.1. (continued).

Unit	SI equivalent
meganewton, MN	1 MN
meganewton metre, MN m	1 MN m
meganewton/square metre, MN/m^2	1 MN m^{-2}
megawatt, MW	1 MW
metre, m	1 m
metre/second, m/s	1 m s^{-1}
metre/second squared, m/s^2	1 m s^{-2}
metre squared/second, m^2/s	1 m^2 s^{-1}
metric horsepower	735.49 W
mho	1 Ω^{-1}
microbar, μbar or μb	0.1 N m^{-2}
microgram, μg	1 μg
microinch, μin	0.0254 μm
micrometre, μm	1 μm
micrometre of mercury, μmHg	0.133 322 N m^{-2}
micron, μm	1 μm
micronewton metre, μN m	1 μN m
micronewton/square metre, μN/m^2	1 μN m^{-2}
microsecond, μs	1 μs
microwatt, μW	1 μW
mil	25.4 μm
mile	1.609 34 km
mile, international nautical n mile	1.852 km
mile, nautical U.K.	1.853 184 km
mile, nautical/hour (knot) U.K.	1.853 184 km h^{-1}
mile, telegraph nautical	1.855 32 km
mile/gallon, mile/gal	0.354 006 km dm^{-3}
mile/hour, mile/h	1.609 34 km h^{-1}
mile/U.S. gallon, mile/U.S. gal	0.425 144 km dm^{-3}
millibar, mbar or mb	100 N m^{-2}
milligal, mGal or mgal	10^{-5} m s^{-2}
milligram, mg	1 mg
milli-inch ("thou")	25.4 μm
millijoule, mJ	1 mJ
millilitre, ml	10^{-6} m^3
millimetre, mm	1 mm
millimetre of mercury, mmHg	133.322 N m^{-2}
millimetre of water	9.806 65 N m^{-2}
millinewton, mN	1 mN
millinewton/metre, mN/m	1 mN m^{-1}
millinewton/square metre, mN/m^2	1 mN m^{-2}
millionth of an inch, μin.	0.0254 m
milliradian, mrad	1 mrad
millisecond, ms	1 ms
millitorr	0.133 322 N m^{-2}

TABLE 5.1. (continued).

Unit	SI equivalent
milliwatt, mW	1 mW
minim	59.193 9 mm^3
minute (angle), '	$\pi/180 \times 60$ rad $= \pi/10\ 800$ rad
minute (time), min	60 s
molality	1 mol kg^{-1}
molar, mol/l	1 mol dm^{-3}
mole, mol	1 mol
nanometre, nm	1 nm
nanosecond, ns	1 ns
nautical mile (U.K.)	1.853 184 km
nautical mile (telegraph)	1.855 32 km
newton, N	1 N
newton metre, N m	1 N m
newton/metre, N/m	1 N m^{-1}
newton/square metre, N/m^2	1 N m^{-2}
newton/square millimetre, N/mm^2	1 N mm^{-2}
newton second/metre squared, N s/m^2	1 N s m^{-2}
nit	1 cd m^{-2}
oersted, Oe	$10^3/4\pi$ A m^{-1}
ohm, Ω	1 Ω
ounce, oz	0.028 3495 kg
ounce, apothecaries'; oz apoth	0.031 1035 kg
ounce, fluid, fl oz	28.4131 cm^3
ounce-force, ozf	0.278 014 N
ounce-force inch, ozf-in.	7061.55 μN m
ounce/gallon, oz/gal	6.236 02 kg m^{-3}
ounce/inch, oz/in.	1.116 12 kg m^{-1}
ounce inch squared, oz in^2	0.182 900 $\times$ 10^{-4} kg m^2
ounce/square foot, oz/ft^2	0.305 152 kg m^{-2}
ounce/square yard, oz/yd^2	0.033 9057 kg m^{-2}
ounce, troy, oz tr	0.301 1035 kg
parsec	30.87 $\times$ 10^{15}m
pascal, Pa	1 N m^{-2}
peck	9.092 18 dm^3
pennyweight, dwt	1.555 17 g
perch	5.029 2 m
petrograd standard (timber)	4.672 28 m^3
pieze	1000 N m^{-2}
pint, pt	0.568 261 dm^3
poise, P	0.1 kg m^{-1} s^{-1}
poiseuille, Pl	1 N s m^{-2}
pole	5.0292 m
pound, lb	0.453592 37 kg

TABLE 5.1. (continued).

Unit	SI equivalent
poundal, pdl	0.138 255 N
poundal foot, pdl ft	0.042 140 1 N m
poundal second/square foot, pdl s/ft²	1.488 16 N s m^{-2}
poundal/square foot, pdl/ft²	1.488 16 N m^{-2}
pound/cubic foot, lb/ft³	16.0185 kg m^{-3}
pound/cubic inch, lb/in³	27.6799 Mg m^{-3}
pound/foot, lb/ft	1.488 16 kg m^{-1}
pound foot/second, lb ft/s	0.138 255 kg m s^{-1}
pound/foot second, lb/ft s	1.488 16 kg m^{-1} s^{-1}
pound foot squared, lb ft²	0.042 1401 kg m²
pound foot squared/second, lb ft²/s	0.042 1401 kg m² s^{-1}
pound force, lbf	4.448 22 N
pound-force foot, lbf-ft	1.355 82 N m
pound-force/foot, lbf/ft	14.5939 N m^{-1}
pound-force hour/square foot, lbf h/ft²	0.172 369 MN s m^{-2}
pound-force inch, lbf in.	0.112 985 N m
pound-force second/square foot, lbf s/ft²	47.8803 N s m^{-2}
pound-force/square foot, lbf/ft²	47.8803 N m^{-2}
pound-force/square inch, lbf/in² or psi	6894.76 N m^{-2}
pound/gallon, lb/gal	0.099 7763 Mg m^{-3}
pound/inch, lb/in.	17.8580 kg m^{-1}
pound inch squared, lb in²	2.926 40 $\times$ 10^{-4} kg m²
pound/second, lb/s	0.453 592 kg s^{-1}
pound/square foot, lb/ft²	4.882 43 kg m^{-2}
pound/square foot inch, lb/ft² in	192.222 kg m^{-3}
pound/square inch, lb/in²	703.070 kg m^{-2}
pound/square inch foot, lb/in² ft	2306.66 kg m^{-3}
pound/1000 square feet, lb/1000 ft²	4.882 43 g m^{-2}
pound/yard, lb/yd	0.496 055 kg m^{-1}
psi	6894.76 N m^{-2}
quart, qt	1.136 52 dm³
quarter	12.7006 kg
quintal	100 kg
rad (100 ergs/g)	0.01 J kg^{-1}
radian, rad	1 rad
radian/second, rad/s	1 rad s^{-1}
radian/second squared, rad/s²	1 rad s^{-2}
Rankine degree, °R	*see* degree Rankine
revolution/minute, rev/min	1/60 s^{-1}
revolution/second, rev/s	1 s^{-1}
rod	5.0292 m
rood	1011.71 m²
scruple	1.295 98 g
second (angle), ''	$\pi/180 \times 60 \times 60$ rad $= \pi/648\,000$ rad
second (time), s	1 s
siemens, S	1 Ω^{-1}
slug	14.5939 kg
slug/cubic foot, slug/ft³	515.379 kg m^{-3}
slug/foot second, slug/ft s	47.8803 kg m^{-1} s^{-1}

TABLE 5.1. (continued).

Unit	SI equivalent
slug foot squared, slug ft²	1.355 82 kg m²
square centimetre, cm²	10^{-4} m²
square chain	404.686 m²
square foot, ft²	0.092 9030 m²
square foot hour degree F/British thermal unit foot, ft² h °F/Btu ft	0.577 789 m K W⁻¹
square foot hour degree F/British thermal unit inch, ft² h °F/Btu in.	6.933 47 mK W⁻¹
square inch, in²	645.16 mm²
square kilometre, km²	1 km²
square metre, m²	1 m²
square mile, sq. mile	2.589 99 km²
square mile/ton, sq. mile/ton	2549.08 m² kg⁻¹
square millimetre, mm²	1 mm²
square yard, yd²	0.836 127 m²
square yard/ton, yd²/ton	0.822 922 m² Mg⁻¹
standard, Petrograd (timber)	4.672 28 m³
statcoulomb	$10/c_0$C = 3.335 640 × 10^{-8} C
steradian, sr	1 sr
sthene	1000 N
stoke, St.	10^{-4} m² s⁻¹
stone	6.350 29 kg
tesla, T	1 T
tex (for textile and fibres)	1 Tex
therm	105.506 MJ
therm/gallon, therm/gal	23.2080 GJ m⁻³
thermie, th	4.1855 MJ
thou	25.4 μm
ton	1016.05 kg
ton/cubic yard, ton/yd³	1328.94 kg m⁻³
ton-force, tonf	9.964 02 kN
ton-force foot, tonf ft	3.037 03 kN m
ton-force/foot, tonf/ft	32.6903 kN m⁻¹
ton-force/square foot, tonf/ft²	107.252 kN m⁻²
ton-force/square inch, tonf/in²	15.4443 MN m⁻²
ton mile	1635.17 kg km
ton/mile	0.631 342 kg m⁻¹
ton of refrigeration	3516.85 W
ton/square mile, ton/sq mile	392.298 kg km⁻²
ton/1000 yards, ton/1000 yd	1.111 16 kg m⁻¹
tonne (metric ton), t	1 Mg
torr	133.322 N m⁻²
torr litre/second, torr l/s	0.133 22 N m s⁻¹
unified atomic mass unit	1.6604 × 10^{-27} kg
volt, V	1 V
volt/metre, V/m	1 V m⁻¹
volt/mil, V/mil	39.3701 kV m⁻¹

TABLE 5.1. (continued).

Unit	SI equivalent
watt, W	1 W
watt/metre degreecelsius, W/m °C	$1\ \text{W m}^{-1}\ \text{K}^{-1}$
watt/square metre, W/m²	$1\ \text{W m}^{-2}$
watt/square metre degree celsius, W/m² °C	$1\ \text{W m}^{-2}\ \text{K}^{-1}$
weber, Wb	1 Wb
weber/square metre, Wb/m²	$1\ \text{Wb m}^{-2}$
yard, yd	0.9144 m
yard/pound, yd/lb	$2.015\ 91\ \text{m kg}^{-1}$

CHAPTER 6

Engineering Data in SI Units
(Direct reading charts and tables)

TABLE 6.1. Sectional Properties of Metals in SI Units

Substance	E, N/mm^2	G, N/mm^2	σ	K, N/mm^2	Tensile strength, N/mm^2
Aluminium	70 300	26 100	0.345	75 500	90 − 150
Brass	101 000	37 300	0.350	111 800	280 − 730
Copper	129 800	48 300	0.343	137 800	120 − 400
Iron (cast)	152 000	60 000	0.270	109 000	100 − 230
Iron (wrought)	211 400	81 000	0.293	170 000	260 − 450
Lead	16 100	5 600	0.440	45 700	12 − 17
Magnesium	44 700	17 000	0.291	25 600	60 − 190
Silver	82 700	30 200	0.366	103 600	300
Platinum	168 000	61 000	0.377	228 000	330 − 370
Tantalum	185 700	69 200	0.342	196 300	800 − 1100
Tin	49 900	18 400	0.357	58 200	20 − 35
Tungsten	411 000	160 000	0.280	311 500	1500 − 3500
Steel (mild)	211 980	82 200	0.291	169 200	430 − 690
Steel (hardened)	201 400	77 800	0.295	165 200	1800 − 2300

Note: 1 N/mm^2 = 1 MN/m^2 = 1 × 10^6 N/m^2.
    E − Young's modulus (or modulus of elasticity).
    G − Shear or rigidity modulus.
    K − Bulk modulus.
    σ − Poisson's ratio.

TABLE 6.2. Physical Properties of Important Moulded Plastics

Material	Type of plastic	Colour	Tensile strength, σ = MN/m²	Young's modulus (E), E = GN/m²	Relative density, d	Water absorption, %	Static coefficient of friction on steel, μ
Cellulose acetate (sheet)	Thermoplastic	All colours: transparent and opaque	41 – 76	14 – 38	1.2 – 1.4	1.5 – 3.0	–
Cellulose nitrate	Thermoplastic	All colours: transparent and opaque	34 – 68	1.4 – 2.8	1.3 – 1.6	1.0 – 3.0	–
Laminated phenolic (asbestos filled)	Thermosetting	Natural	48 – 117	2.4 – 10.3	1.55 – 1.80	0.3 – 2.0	Unlubricated, μ = 0.22 Lubricated, μ = 0.10
Nylon	Thermoplastic	Various	80 upwards	2.75	1.14	1.5	Lubricated, μ = 0.20
Polytetrafluorethylene Teflon or PTFE	Thermosetting	Natural	10.3 – 20.6	–	2.2	–	Unlubricated, μ = 0.04
Polythene	Thermoplastic	Various	9 – 12	–	0.92	0	–
Urea formaldehyde (wood filled)	Thermosetting	Unlimited shades	55 – 76	7.5 – 9.6	1.5	1.0 – 1.5	–

TABLE 6.3. Some Important Physical Constants expressed in SI Units

Description	Symbol	Numerical value in SI units
Avogadro's number	N	$= 6.023 \times 10^{26}$ kg mol
Bohr magneton	β	$= 9.27 \times 10^{-24}$ A m^2
Boltzmann's constant	k	$= 1.380 \times 10^{-23}$ J/deg K
Characteristic impedance of free space	Z_0	$= (\mu_0/E_0)^{1/2} = 120\,\pi\Omega$
Electron volt	eV	$= 1.602 \times 10^{-19}$ J
Electric charge	e	$= 1.602 \times 10^{-19}$ C
Electronic rest mass	m_e	$= 9.109 \times 10^{-31}$ kg
Electronic charge to mass ratio	e/m_e	$= 1.759 \times 10^{11}$ C/kg
Energy for $T = 290°$K	kT	$= 4 \times 10^{-21}$ J
Energy for ground state atom (Rydberg energy)	H	$= 13.60$ eV
Faraday constant	F	$= 9.65 \times 10^7$ C/kg mol
Permeability of free space	μ_0	$= 4\pi \times 10^{-7}$ H/m
Permittivity of free space	E_0	$= \dfrac{1}{36\,\pi} \times 10^{-9}$ F/m
Planck's constant	h	$= 6.626 \times 10^{-34}$ J s
Proton mass	m_p	$= 1.672 \times 10^{-27}$ kg
Proton to electron mass ratio	m_p/m_e	$= 1836.1$
Radius of first H orbit (Bohr atom)		$= 0.529 \times 10^{-10}$ m $= 0.529$ Å
Standard gravitational acceleration	g	$= 9.807$ m/s^2
Stefan Boltzmann constant	σ	$= 5.67 \times 10^{-8}$ J/m^2 s deg K^4
Universal constant of gravitation	G	$= 6.67 \times 10^{-11}$ N m^2/kg^2
Universal gas constant	R	$= 8.314$ kJ/kg mol deg K
Velocity of light *in vacuo*	c	$= 2.9979 \times 10^8$ m/s
Volume of 1 kg mol of ideal gas at n.t.p.		$= 22.42$ m^3

TABLE 6.4. Some Important Non-SI Units

The following non-SI units would continue to be used for a very long time:

Quantity	SI unit	Associated non-SI units
Angle	rad	degree °, minute ′, second ″
Time	s	hour (h); minute (min)
Mass	kg	Tonne (t) $= 10^3$ kg $= 1$ Mg
Area	m^2	hectare (ha) $= 10^4$ m^2
Volume	m^3	litre (l.) $= 10^{-3}$ m^3
Pressure	Pa	bar (bar) $= 10^5$ N/m^2 $= 10^5$ Pa
Energy	J	kWh $= 3.6$ MJ
Stress	Pa; N/m^2	hectobar (hbar) $= 10^7$ N/m^2 $= 10^7$ Pa

TABLE 6.5. Temperature Conversion
(Range of temperature covered 0 to 300)

For any reading to be converted from °F to °C (or vice versa) locate the reading in the middle column indicating $\frac{C}{F}$, the equivalent reading is obtained on the same horizontal line towards the right or left as the case may be:

C	$\frac{C}{F}$	F	C	$\frac{C}{F}$	F
−17.8	0	32	10.0	50	122.0
−17.2	1	33.8	10.6	51	123.8
−16.7	2	35.6	11.1	52	125.6
−16.1	3	37.4	11.7	53	127.4
−15.6	4	39.2	12.2	54	129.2
−15.0	5	41.0	12.8	55	131.0
−14.4	6	42.8	13.3	56	132.8
−13.9	7	44.6	13.9	57	134.6
−13.3	8	46.4	14.4	58	136.4
−12.8	9	48.2	15.0	59	138.2
−12.2	10	50.0	15.6	60	140.0
−11.7	11	51.8	16.1	61	141.8
−11.1	12	53.6	16.7	62	143.6
−10.6	13	55.4	17.2	63	145.4
−10.0	14	57.2	17.8	64	147.2
−9.4	15	59.0	18.3	65	149.0
−8.9	16	60.8	18.9	66	150.8
−8.3	17	62.6	19.4	67	152.6
−7.8	18	64.4	20.0	68	154.4
−7.2	19	66.2	20.6	69	156.2
−6.7	20	68.0	21.1	70	158.0
−6.1	21	69.8	21.7	71	159.8
−5.6	22	71.6	22.2	72	161.6
−5.0	23	73.4	22.8	73	163.4
−4.4	24	75.2	23.3	74	165.2
−3.9	25	77.0	23.9	75	167.0
−3.3	26	78.8	24.4	76	168.8
−2.8	27	80.6	25.0	77	170.6
−2.2	28	82.4	25.6	78	172.4
−1.7	29	84.2	26.1	79	174.2
−1.1	30	86.0	26.7	80	176.0
−0.6	31	87.8	27.2	81	177.8
0.0	32	89.6	27.8	82	179.6
0.6	33	91.4	28.3	83	181.4
1.1	34	93.2	28.9	84	183.2
1.7	35	95.0	29.4	85	185.0
2.2	36	96.8	30.0	86	186.8
2.8	37	98.6	30.6	87	188.6
3.3	38	100.4	31.1	88	190.4
3.9	39	102.2	31.7	89	192.2
4.4	40	104.0	32.2	90	194.0
5.0	41	105.8	32.8	91	195.8
5.6	42	107.6	33.3	92	197.6
6.1	43	109.4	33.9	93	199.4
6.7	44	111.2	34.4	94	201.2
7.2	45	113.0	35.0	95	203.0
7.8	46	114.8	35.6	96	204.8
8.3	47	116.6	36.1	97	206.6
8.9	48	118.4	36.7	98	208.4
9.4	49	120.2	37.2	99	210.2

TABLE 6.5. (continued).

C	C F	F	C	C F	F
38	100	212	99	210	410
43	110	230	100	212	413.6
49	120	248	104	220	428
54	130	266	110	230	446
60	140	284	116	240	464
66	150	302	121	250	482
71	160	320	127	260	500
77	170	338	132	270	518
82	180	356	138	280	536
88	190	374	143	290	554
93	200	392	149	300	572

Example (i): $59°F = ?°C$
Look up in column $\frac{C}{F}$
Answer is to the left of 59, i.e. 15°C

Example (ii): $75°C = ?°F$
Look up 75 in column $\frac{C}{F}$
Answer is in column F to the right of this figure, i.e. 167°F

TABLE 6.6. Length Conversion
(Inches and millimetres)

For any length dimension to be converted from inches to millimetres (or vice versa) locate the reading in the middle column indicating mm; in., equivalent reading is obtained on the same horizontal line towards the right or the left as the case may be.

mm	mm; in	in.	mm	mm; in	in.
0.254	0.01	0.000 39	15.24	0.6	0.023 62
0.508	0.02	0.000 79	17.78	0.7	0.027 56
0.762	0.03	0.001 18	20.32	0.8	0.031 50
1.016	0.04	0.001 57	22.86	0.9	0.035 43
1.270	0.05	0.001 97	25.40	1.0	0.039 4
1.524	0.06	0.002 36	50.80	2	0.078 7
1.778	0.07	0.002 76	76.20	3	0.118 1
2.032	0.08	0.003 15	101.6	4	0.157 5
2.286	0.09	0.003 54	127.0	5	0.196 9
2.54	0.10	0.003 94	152.4	6	0.236 2
5.08	0.20	0.007 87	177.8	7	0.275 6
7.62	0.3	0.011 81	203.2	8	0.315 0
10.16	0.4	0.015 75	228.6	9	0.354 3
12.70	0.5	0.019 69	254.0	10	0.393 7

Example (i);   9.57 in. = ? mm   *Example* (ii);   8.64 mm = ? in.

　　　　9 in. = 228.6 mm　　　　　　　　　　　8 mm = 0.3150 in.
　　0.5 in. =   12.7 mm　　　　　　　　0.6 mm = 0.02362 in.
　0.07 in. =   1.778 mm　　　　　　0.04 mm = 0.00157 in.

　　9.57 in. = 243.078 mm　　　　　　　8.64 mm = 0.34019 in.

TABLE 6.7. Decimal Equivalents for Fractions of an Inch

Fractions of an inch	Decimals of an inch	mm	Fractions of an inch	Decimals of an inch	mm
1/64	0.015 625	0.397	33/64	0.515 625	13.096
1/32	0.03125	0.794	17/32	0.531 25	13.494
3/64	0.046 875	1.191	35/64	0.546 875	13.891
1/16	0.0625	1.588	9/16	0.5625	14.288
5/64	0.078 125	1.984	37/64	0.578 125	14.684
3/32	0.093 75	2.381	19/32	0.593 75	15.081
7/64	0.109 375	2.778	39/64	0.609 375	15.478
1/8	0.125	3.175	5/8	0.625	15.875
9/64	0.140 625	3.572	41/64	0.640 625	16.272
5/32	0.156 25	3.969	21/32	0.656 25	16.669
11/64	0.171 875	4.366	43/64	0.671 875	17.066
3/16	0.1875	4.763	11/16	0.6875	17.463
13/64	0.203 125	5.159	45/64	0.703 125	17.859
7/32	0.218 75	5.556	23/32	0.718 75	18.256
15/64	0.234 375	5.953	47/64	0.734 375	18.653
1/4	0.25	6.350	3/4	0.75	19.050
17/64	0.265 625	6.747	49/64	0.765 625	19.447
9/32	0.281 25	7.144	25/32	0.781 25	19.844
19/64	0.296 875	7.541	51/64	0.796 875	20.241
5/16	0.3125	7.938	13/16	0.8125	20.638
21/64	0.328 125	8.334	53/64	0.828 125	21.034
11/32	0.343 75	8.731	27/32	0.843 75	21.431
23/64	0.359 375	9.128	55/64	0.859 375	21.828
3/8	0.375	9.525	7/8	0.875	22.225
25/64	0.390 625	9.922	57/64	0.890 625	22.622
13/32	0.406 25	10.319	29/32	0.906 25	23.019
27/64	0.421 875	10.716	59/64	0.921 875	23.416
7/16	0.4375	11.113	15/16	0.937 5	23.813
29/64	0.453 125	11.509	61/64	0.953 125	24.209
15/32	0.468 75	11.906	31/32	0.968 75	24.606
31/64	0.484 375	12.303	63/64	0.984 375	25.003
1/2	0.5	12.700	1	1.00	25.400

TABLE 6.8. Length Conversion
(Inches and decimal of an inch into millimetres)

Inches	0	1	2	3	4	5	6	7	8	9
					Millimetres (mm)					
0	0	25.4	50.8	76.2	101.6	127.0	152.4	177.8	203.2	228.6
10	254.0	279.4	304.8	330.2	355.6	381.0	406.4	431.8	457.2	482.6
20	508.0	553.4	558.8	584.2	609.6	635.0	660.4	685.8	711.2	736.6
30	762.0	787.4	812.8	838.2	863.6	889.0	914.4	939.8	965.2	990.6
40	1016.0	1041.4	1066.8	1092.2	1117.6	1143.0	1168.4	1193.8	1219.2	1244.6
50	1270.0	1295.4	1320.8	1346.2	1371.6	1397.0	1422.4	1447.8	1473.2	1498.6
60	1524.0	1549.4	1574.8	1600.2	1625.6	1651.0	1676.4	1701.8	1727.2	1752.6
70	1778.0	1803.4	1828.8	1854.2	1879.6	1905.0	1930.4	1955.8	1981.2	2006.6
80	2032.0	2057.4	2082.8	2108.2	2133.6	2159.0	2184.4	2209.8	2235.2	2260.6
90	2286.0	2311.4	2336.8	2362.2	2387.6	2413.0	2430.4	2463.8	2489.2	2514.6
100	2450.0									

Auxiliary Table for decimals of an inch

Inches	0.1	0.2	0.3	0.4	0.5	0.6	0.7	0.8	0.9
Inches	—								
Millimetres	2.54	5.08	7.62	10.16	12.70	15.24	17.78	20.32	22.86

Example (i): 93 in. ? mm
93 in. = 2362.2 mm (from table)

Example (ii): 27.4 in. = ? mm.   27 in. = 685.8 mm (from table)
0.4 in. = 10.16 mm (from table)

27.4 in. = 695.24 mm

TABLE 6.9. Length Conversion
(Miles to kilometres)

Miles	0	1	2	3	4	5	6	7	8	9
					Kilometres (km)					
0	—	1.61	3.22	4.83	6.44	8.05	9.66	11.27	12.87	14.48
10	16.09	17.70	19.31	20.92	22.53	24.14	25.75	27.36	28.97	30.58
20	32.19	33.80	35.41	37.01	38.62	40.23	41.84	43.45	45.06	46.67
30	48.28	49.89	51.50	53.11	54.72	56.33	57.94	59.55	61.16	62.67
40	64.37	65.98	67.59	69.20	70.81	72.42	74.03	75.64	77.25	78.86
50	80.47	82.08	83.69	85.30	86.90	88.51	90.12	91.73	93.34	94.95
60	96.56	98.17	99.78	101.39	103.00	104.61	106.22	107.83	109.44	111.05
70	112.65	114.26	115.87	117.48	119.09	120.70	122.31	123.92	125.53	127.14
80	128.75	130.36	131.97	133.58	135.19	136.79	138.40	140.01	141.62	143.23
90	144.84	146.45	148.06	149.67	151.28	152.89	154.50	156.11	157.72	159.33
100	160.94	—	—	—	—	—	—	—	—	—

Auxiliary tables (fractions of miles)

Miles	0.1	0.2	0.3	0.4	0.5	0.6	0.7	0.8	0.9	1.0
km	0.161	0.322	0.483	0.644	0.805	0.966	1.127	1.288	1.448	1.609

Furlongs	1	2	3	4	5	6	7	8
km	0.201	0.402	0.604	0.804	1.006	1.207	1.408	1.609

1 mile = 1.609 34 km: 1 furlong = 0.201 17 km

Example (i): 94.7 miles = ? km

$$94 \text{ miles} = 151.28 \text{ km} \quad \text{(from table)}$$
$$0.7 \text{ miles} = 1.127 \text{ km} \quad \text{(from Auxiliary tables)}$$
$$94.7 \text{ miles} = 152.407 \text{ km}$$

Example (ii): 37 miles 3 furlongs = ? km

$$37 \text{ miles} = 59.55 \text{ km} \quad \text{(from tables)}$$
$$3 \text{ furlongs} = 0.604 \text{ km} \quad \text{(from Auxiliary tables)}$$
$$37 \text{ miles 3 furlongs} = 60.154 \text{ km}$$

TABLE 6.10. Area Conversion
(Square inches and square metres)

For any area to be converted from in^2 to m^2 (and vice versa) locate the reading in the middle column of this table indicating $\frac{m^2}{in^2}$, the equivalent reading is obtained on the place towards right or the left of this figure as the case may be.

in^2	$\frac{m^2}{in^2}$	m^2	in^2	$\frac{m^2}{in^2}$	m^2
15.5	0.01	0.000 006 425	3100	2	0.001 290
31.0	0.02	0.000 012 9	4650	3	0.001 935
46.5	0.03	0.000 019 35	6200	4	0.002 581
62.0	0.04	0.000 025 81	7750	5	0.003 226
77.5	0.05	0.000 032 26	9300	6	0.003 871
93.0	0.06	0.000 038 71	10 850	7	0.004 516
108.5	0.07	0.000 045 16	12 400	8	0.005 161
124.0	0.08	0.000 051 61	13 950	9	0.005 806
139.5	0.9	0.000 058 06	15 500	10	0.006 452
155	0.1	0.000 064 52	31 000	20	0.012 9
310	0.2	0.000 129 0	46 500	30	0.019 35
465	0.3	0.000 193 5	62 000	40	0.025 81
620	0.4	0.000 258 1	77 500	50	0.032 26
775	0.5	0.000 322 6	93 000	60	0.038 71
930	0.6	0.000 387 1	108 500	70	0.045 16
1085	0.7	0.000 451 6	124 000	80	0.051 61
1240	0.8	0.000 516 1	139 500	90	0.058 06
1395	0.9	0.000 580 6	155 000	100	0.064 52
1550	1	0.000 645 2			

Example (i):   97.54 in^2 = ? m^2

90 in^2 = 0.058 06
7 in^2 = 0.004 516
0.5 in^2 = 0.000 322 6
0.04 in^2 = 0.000 025 81
97.54 in^2 = 0.062 924 41 m^2

Example (ii):   5.3 m^2 = ? in^2

5 m^2 = 7750
0.3 m^2 = 465
5.3 m^2 = 8215 in^2

TABLE 6.11. Volume Conversion
(cubic inch and cubic metre)

For a volume in in^3 to be converted to m^3 (or vice versa) locate the figure in the middle column, i.e. $\frac{m^3}{in^3}$, and the desired conversion is read to the right or left as the case may be.

m^3	$\frac{m^3}{in^3}$	in^3
0.000 016 39	1	61 020
0.000 032 77	2	122 050
0.000 049 16	3	183 070
0.000 065 55	4	244 100
0.000 081 94	5	305 100
0.000 098 32	6	366 100
0.000 114 7	7	427 200
0.000 131 1	8	488 200
0.000 147 5	9	549 200
0.000 163 9	10	610 200

Example (i):   4 in^3 = ? m^3
= 0.000 065 55 m^3

Example (ii): 7 m^3 = ? in^3
= 427 200 in^3

TABLE 6.12. Capacity
(Imperial gallons to litres)

Gallon	0	1	2	3	4	5	6	7	8	9
					Litres					
0	–	4.55	9.09	13.64	18.18	22.73	27.28	31.82	36.37	40.91
10	45.46	50.01	54.55	59.10	63.65	68.19	72.74	77.28	81.83	86.38
20	90.92	95.47	100.01	104.56	109.11	113.65	118.20	122.74	127.29	131.84
30	136.38	140.93	145.48	150.02	154.57	159.11	163.66	168.21	172.75	177.30
40	181.84	186.39	190.94	195.48	200.03	204.57	209.12	213.67	218.21	222.76
50	227.31	231.85	236.40	240.94	245.49	250.04	254.58	259.13	263.67	268.22
60	273.77	277.31	281.86	286.40	290.95	295.50	300.04	304.59	309.13	313.68
70	318.23	322.77	327.32	331.87	336.41	340.96	345.50	350.05	354.60	359.14
80	363.69	368.23	372.78	377.33	381.87	386.42	390.96	395.51	400.06	404.60
90	409.15	413.69	418.24	422.79	427.33	431.88	436.43	440.97	445.52	450.06
100	454.61	–	–	–	–	–	–	–	–	–

Auxiliary tables for fractions of gallon

Pint	1	2	3	4	5	6	7	8
Litres	0.568	1.136	1.705	2.273	2.841	3.410	3.978	4.546

Conversion factors: 1 gallon = 4.5461 dm^3 = 4.5461 litres.
                     1 pint   = 0.568 26 dm^3 = 0.568 26 litres.

Example: 65 gallons 2 pints = ? litres
65 gallons = 295.50 litres (from table)
2 pints = 1.136 litres (from Auxiliary table)
---
65 gallons 2 pints = 296.636 litres

TABLE 6.13. Mass Conversion
(Pounds to kilograms)

Mass in Pounds (lb)	0	1	2	3	4	5	6	7	8	9
					Kilograms (k_g)					
0	0.00	0.454	0.907	1.361	1.814	2.268	2.722	3.175	3.629	4.082
10	4.536	4.990	5.443	5.897	6.350	6.804	7.257	7.711	8.165	8.618
20	9.072	9.525	9.979	10.433	10.886	11.340	11.793	12.247	12.701	13.154
30	13.608	14.061	14.515	14.969	15.422	15.876	16.329	16.783	17.237	17.690
40	18.144	18.597	19.051	19.505	19.958	20.412	20.865	21.319	21.772	22.226
50	22.680	23.133	23.587	24.040	24.494	24.948	25.401	25.855	26.308	26.762
60	27.216	27.669	28.123	28.576	29.030	29.484	29.937	30.391	30.844	31.298
70	31.752	32.205	32.659	33.112	33.566	34.019	34.473	34.927	35.380	35.834
80	36.287	36.741	37.195	37.648	38.102	38.555	39.009	39.463	39.916	40.370
90	40.823	41.277	41.731	42.184	42.638	43.091	43.549	43.999	44.452	44.906
100	45.359	—	—	—	—	—	—	—	—	—

Auxiliary table for fractions

lb	0.1	0.2	0.3	0.4	0.5	0.6	0.7	0.8	0.9
kg	0.045	0.091	0.136	0.181	0.227	0.272	0.317	0.363	0.408

1 lb = 0.453 59 kg

Example (i): 37 lb = ? kg
= 16.783 kg (from table)

Example (ii): 85.6 lb = ? kg
85 lb = 38.555 kg (from table)
0.6 lb = 0.272 kg (from Auxiliary tables)
85.6 lb = 38.827 kg

TABLE 6.14. Pressure Conversion

Inches of water	Inches of mercury	lb/in^2	Pa
1	0.074	0.036	250
2	0.148	0.072	500
3	0.221	0.108	750
4	0.295	0.144	1000
5	0.369	0.180	1250
6	0.443	0.216	1500
7	0.517	0.252	1750
8	0.590	0.288	2000
9	0.664	0.324	2250
10	0.738	0.360	2500
11	0.812	0.396	2750
12	0.886	0.432	3000
13	0.959	0.468	3250
14	1.033	0.504	3500
15	1.107	0.540	3750
16	1.180	0.576	4000
17	1.255	0.612	4250
18	1.328	0.648	4500
19	1.402	0.684	4750
20	1.476	0.720	5000
21	1.550	0.756	5250
22	1.624	0.792	5500
23	1.697	0.828	5750
24	1.771	0.864	6000
25	1.845	0.900	6250
26	1.919	0.936	6500
27	1.993	0.972	6750
28	2.066	1.008	7000
29	2.140	1.044	7250
30	2.214	1.080	7500
31	2.288	1.116	7750
32	2.362	1.152	8000
33	2.435	1.188	8250
34	2.509	1.244	8500
35	2.583	1.260	8750

TABLE 6.15. Stress Conversion
[lb/in² (psi) to newtons per square millimetre (N/mm²)]

lb per square in. (lb/in²)	0	1	2	3	4	5	6	7	8	9
	Newtons per square millimetre (N/mm²) × 10⁻³									
0	0.00	6.89	13.79	20.68	27.58	34.47	41.37	48.26	55.16	62.05
10	68.95	75.84	82.74	89.63	96.53	103.42	110.32	117.21	124.11	131.00
20	137.90	144.79	151.68	158.58	165.47	172.37	179.26	186.16	193.05	199.95
30	206.84	213.74	220.63	227.53	234.42	241.32	248.21	255.11	262.00	268.90
40	275.79	282.69	289.58	296.48	303.37	310.26	317.16	324.05	330.95	337.84
50	344.74	351.63	358.53	365.42	372.32	379.21	386.11	393.00	399.90	406.79
60	413.69	420.58	427.48	434.37	441.27	448.16	455.05	461.95	468.84	475.74
70	482.63	489.53	496.42	503.32	510.21	517.11	524.00	530.90	537.79	544.69
80	551.58	558.48	565.37	572.27	579.16	586.06	592.95	599.84	606.74	613.63
90	620.53	627.42	634.32	641.21	648.11	655.00	661.90	668.79	675.69	682.58
100	689.48	—	—	—	—	—	—	—	—	—

Auxiliary table for decimals

lb/in²	0.1	0.2	0.3	0.4	0.5	0.6	0.7	0.8	0.9
N/mm² × 10⁻³	0.689	1.379	2.068	2.758	3.447	4.137	4.826	5.516	6.205

1 lb/in² = 0.00689476.   N/mm² = 6.894 76 × 10⁻³ N/mm².

Example (i): 94 lb/in² = ? N/mm²
= 648.11 × 10⁻³ N/mm² (from table)

Example (ii): 47.6 lb/in² = ? N/mm²

47 lb/in² = 324.05 (from table)
0.6 lb/in² = 4.137   (from Auxiliary table)
47.6 lb/in² = 328.187 × 10⁻³ N/mm²

TABLE 6.16. Stress Conversion
Ton-force per square inch (t/in²) to newtons per square millimetre (N/mm²)
$$1 \text{ t/in}^2 = 15.4443 \text{ N/mm}^2$$

Ton-force per square inch (t/in²)	0	1	2	3	4	5	6	7	8	9
				Newtons per square millimetre (N/mm²)						
0	0.00	15.44	30.89	46.33	61.78	77.22	92.66	108.12	123.55	138.99
10	154.44	169.88	185.33	200.77	216.22	231.66	247.10	262.55	277.99	293.44
20	308.89	324.33	339.78	355.22	370.67	386.11	401.55	417.00	432.44	447.88
30	463.33	478.77	494.22	509.66	525.11	540.55	555.99	571.44	586.88	602.33
40	617.77	633.21	648.66	664.10	679.55	694.99	710.43	725.88	741.32	756.77
50	772.21	787.66	803.11	818.55	834.00	849.44	864.88	880.33	895.77	911.21
60	926.66	942.10	957.55	972.99	988.44	1003.88	1019.32	1034.77	1050.21	1065.66
70	1081.10	1096.54	1111.99	1127.43	1142.88	1158.32	1173.76	1189.21	1204.65	1220.10
80	1235.54	1250.98	1266.43	1281.87	1297.32	1312.76	1328.20	1343.65	1359.09	1374.54
90	1389.99	1405.43	1420.88	1436.32	1451.77	1467.21	1482.65	1498.10	1513.54	1528.99
100	1544.43	–	–	–	–	–	–	–	–	–

Example (i): 23 t/in² = ? N/mm² (read column under 3 on straight line from 20) = 355.22 N/mm²

Example (ii):

46.3 t/in²	= ? N/mm²	= ? N/mm²
46 t/in²		= 710.43
0.3 t/in²		= 4.633

$$46.3 \text{ t/in}^2 = 715.063 \text{ N/mm}^2$$

TABLE 6.17. h.p./kW Conversion

For a figure in h.p. to be converted to kW (or vice versa) locate the figure in the column $\frac{kW}{h.p.}$ and the desired conversion is read to the right or left of this figure as the case may be.

kW	kW h.p.	h.p.	kW	kW h.p.	h.p.
0.373	0.5	0.67	7.46	10.0	13.40
0.559	0.75	1.00	8.20	11.0	14.71
0.746	1.00	1.34	8.95	12.0	16.10
1.12	1.5	2.03	9.32	12.5	16.62
1.49	2.0	2.68	11.2	15.0	20.1
1.86	2.5	3.35	13.0	17.5	22.4
2.24	3.00	4.02	14.9	20.0	26.8
2.61	3.5	4.68	18.6	25.0	33.5
2.98	4.0	5.36	22.4	30	40.2
3.36	4.5	6.03	26.1	35	46.9
3.73	5.0	6.70	29.8	40	53.6
4.10	5.5	7.36	33.6	45	60.3
4.47	6.0	8.05	37.3	50	67.0
4.85	6.5	8.70	44.7	60	80.5
5.22	7.0	9.39	52.2	70	93.9
5.59	7.5	10.05	55.9	75	100
5.97	8.0	10.70	59.7	80	107
6.34	8.5	11.40	67.1	90	120
6.71	9.0	12.0	74.6	100	134
7.08	9.5	12.72			

Example (i): 7.5 h.p. = ? kW   *Example* (ii): 75 kW = ? h.p.

= 5.59 kW   = 100 h.p.

TABLE 6.18. The Greek Alphabet

Capital letter	Small letter	Name	Capital letter	Small letter	Name
A	α	alpha	N	ν	nu
B	β	beta	Ξ	ξ	xi
Γ	γ	gamma	O	o	omicron
Δ	δ	delta	Π	π	pi
E	ϵ	epsilon	P	ρ	rho
Z	ζ	zeta	Σ	σ	sigma
H	η	eta	T	τ	tau
Θ	θ	theta	Υ	υ	upsilon
I	ι	iota	Φ	ϕ	phi
K	κ	kappa	X	χ	chi
Λ	λ	lambda	Ψ	ψ	psi
M	μ	mu	Ω	ω	omega

TABLE 6.19. Mathematical Constants

e	$= 2.7183$	$\log \pi^2$	$= 0.9943$
$\dfrac{1}{e}$	$= 0.3679$	$\sqrt{\pi}$	$= 1.7725$
$\log_e^{10}$	$= 2.3026$	$\dfrac{\pi}{4}$	$= 0.7854$
$\log_{10}^x$	$= 0.4343 \ln x$		
π	$= 3.1416$	$\sqrt{2}$	$= 1.414$
π^2	$= 9.8696$	$\sqrt{3}$	$= 1.732$
$\dfrac{1}{\pi}$	$= 0.3183$	$\sqrt{5}$	$= 2.236$
$\ln x$	$= 2.3026 \log_{10}^x$	$\sqrt{10}$	$= 3.162$
$\log^\pi$	$= 0.4971$		
		1 rad	$= 57° \ 39'$
		$1°$	$= 0.017\ 45$ rad

TABLE 6.20. Some Standard Values expressed in SI Units

Standard gravitational acceleration $= 9.81$ m/s²

International standard atmosphere (I.S.A.): (Pressure $= 1.013$ bar
$= 101.3$ kPa

Temperature $= 15°C$
$= 288$ K)

Universal gas constant $\bar{R} = 8.314$ kJ/kJ/kg mol K

Molal volume $= \bar{V} = 22.41$ m³/kg mol at ISA pressure and $0°C$.

Composition of air: volumetric analysis: $N_2 = 79\%$, $O_2 = 21\%$
                                           gravitational analysis: $N_2 = 76.7\%$, $O_2 = 23.3\%$

Properties of air:
   Molecular mass $= \bar{M} = 29$
   Specific gas constant $R = 0.287$ kJ/kg K
   Specific heat at constant pressure $C_p = 1.005$ kJ/kg K
   Specific heat at constant volume $C_v = 0.718$ kJ/kg K
   Ratio of specific heats $\gamma = \dfrac{C_p}{C_v} = 1.4$

   Thermal conductivity $k = 0.0253$ W/m K at ISA conditions
   Dynamic viscosity $\mu = 17.9$ μPa s
   Density at ISA conditions $\rho = 1.225$ kg/m³
   Sonic velocity at ISA conditions $a = 340$ m/s

Properties of water:
   Molecular mass $\bar{M} = 18$
   Specific heat at constant pressure at $15°C$ $C_p = 4.186$ kJ/kg K
   Thermal conductivity $k = 595$ kW/m K at $15°C$
   Dynamic viscosity $\mu = 1.14$ mPa s at $15°C$
   Density at $4°C$ $\rho = 1000$ kg/m³

TABLE 6.21. Tex Count Conversion
(For Textile Technologies)

Count (Tex or traditional)	Equivalent count to or from					
	Worsted	Woollen	Galashiels	Cotton	Yd/oz	Metric
1	886	1 940	2 480	591	31 000	1 000
1.5	591	1 290	1 650	394	20 700	667
2	443	969	1 240	295	15 500	500
2.5	354	775	992	236	12 400	400
3	295	646	827	197	10 300	333
3.5	253	554	709	169	8 860	286
4	221	484	620	148	7 750	250
4.5	197	431	551	131	6 890	222
5	177	177	388	496	6 200	200
5.5	161	352	451	107	5 640	182
6	148	323	413	98.4	5 170	167
6.5	136	298	382	90.9	4 770	154
7	127	277	354	84.4	4 430	143
7.5	118	258	331	78.7	4 130	133
8	111	242	310	73.8	3 880	125
8.5	104	228	292	69.5	3 650	118
9	98.4	215	276	65.6	3 440	111
9.5	93.2	204	261	62.2	3 260	105
10	88.6	194	248	59.1	3 100	100
10.5	84.4	185	236	56.2	2 950	95.2
11	80.5	176	225	53.7	2 820	90.9
11.5	77.0	169	216	51.3	2 700	87.0
12	73.8	162	207	49.2	2 580	83.3
12.5	70.9	155	198	47.2	2 480	80.0
13	68.1	149	191	45.4	2 380	76.9
13.5	65.6	144	184	43.7	2 300	74.1
14	63.3	138	177	42.2	2 210	71.4
14.5	61.1	134	171	40.7	2 140	69.0
15	59.1	129	165	39.4	2 070	66.7
15.5	57.1	125	160	38.1	2 000	64.5
16	55.4	121	155	36.9	1 940	62.5
16.5	53.7	117	150	35.8	1 880	60.6
17	52.1	114	146	34.7	1 820	58.8
17.5	50.6	111	142	33.7	1 770	57.1
18	49.2	108	138	32.8	1 720	55.6
18.5	47.9	105	134	31.9	1 680	54.1
19	46.6	102	131	31.1	1 630	52.6
19.5	45.4	99.4	127	30.3	1 590	51.3
20	44.3	96.9	124	29.5	1 550	50.0
20.5	43.2	94.5	121	28.8	1 510	48.8
21	42.2	92.3	118	28.1	1 480	47.6
21.5	41.2	90.1	115	27.5	1 440	46.5
22	40.3	88.1	113	26.8	1 410	45.5
22.5	39.4	86.1	110	26.2	1 380	44.4
23	38.5	84.3	108	25.7	1 350	43.5
23.5	37.7	82.5	106	25.1	1 320	42.6
24	36.9	80.8	103	24.6	1 290	41.7
24.5	36.2	79.1	101	24.1	1 270	40.8
25	35.4	77.5	99.2	23.6	1 240	40.0
25.5	34.7	76.0	97.3	23.2	1 220	39.2

TABLE 6.21. Tex Count Conversion
(For Textile Technologists)

Count (Tex or traditional)	Worsted	Woollen	Galashiels	Cotton	Yd/oz	Metric
			Equivalent count to or from			
26	34.1	74.5	95.4	22.7	1 190	38.5
27	32.8	71.8	91.9	21.9	1 150	37.0
28	31.6	69.2	88.6	21.1	1 110	35.7
29	30.5	66.8	85.5	20.4	1 070	34.5
30	29.5	64.6	82.7	19.7	1 030	33.3
31	28.6	62.5	80.0	19.0	1 000	32.3
32	27.7	60.6	77.5	18.4	969	31.3
33	26.8	58.7	75.2	17.9	939	30.3
34	26.1	57.0	72.9	17.4	912	29.4
35	25.3	55.4	70.9	16.9	886	28.6
36	24.6	53.8	68.9	16.4	861	27.8
37	23.9	52.4	67.0	16.0	838	27.0
38	23.3	51.0	65.3	15.6	816	26.3
39	22.7	49.7	63.6	15.2	795	25.6
40	22.1	48.4	62.0	14.8	775	25.0
41	21.6	48.4	60.5	14.4	756	24.4
42	21.1	46.1	59.0	14.1	738	23.8
43	20.6	45.1	57.7	13.7	721	23.3
44	20.1	44.0	56.4	13.4	705	22.7
45	19.7	43.1	55.1	13.1	689	22.2
46	19.3	42.1	53.9	12.8	674	21.7
47	18.8	41.2	52.8	12.6	660	21.3
48	18.5	40.4	51.7	12.3	646	20.8
49	18.1	39.6	50.6	12.1	633	20.4
50	17.7	38.8	49.6	11.8	620	20.0
51	17.4	38.0	48.6	11.6	608	19.6
52	17.0	37.3	47.7	11.4	596	19.2
53	16.7	36.6	46.8	11.1	585	18.9
54	16.4	35.9	45.9	10.9	574	18.5
55	16.1	35.2	45.1	10.7	564	18.2
56	15.8	34.6	44.3	10.5	554	17.9
57	15.5	34.0	43.5	10.4	544	17.5
58	15.3	33.4	42.8	10.2	534	17.2
59	15.0	32.8	42.0	10.0	525	16.9
60	14.8	32.3	41.3	9.84	517	16.7
61	14.5	31.8	40.7	9.68	508	16.4
61	14.3	31.3	40.0	9.52	500	16.1
63	14.1	30.8	39.4	9.37	492	15.9
64	13.8	30.3	38.8	9.23	484	15.6
65	13.6	29.8	38.2	9.09	477	15.4
66						
67	13.2	28.9	37.0	8.81	463	14.9
68	13.0	28.5	36.5	8.68	456	14.7
69	12.8	28.1	35.9	8.56	449	14.5
70	12.7	27.7	35.4	8.44	443	14.3
71	12.5	27.3	34.9	8.32	437	14.1
72	12.3	26.9	34.4	8.20	431	13.9
73	12.1	26.5	34.0	8.09	425	13.7
74	12.0	26.2	33.5	7.98	419	13.5
75	11.8	25.8	33.1	7.87	413	13.3

TABLE 6.21. Tex Count Conversion
(For Textile Technologists)

Count (Tex or traditional)	Equivalent count to or from					
	Worsted	Woollen	Galashiels	Cotton	Yd/oz	Metric
76	11.7	25.5	32.6	7.77	408	13.2
77	11.5	25.2	32.2	7.67	403	13.0
78	11.4	24.8	31.8	7.57	397	12.8
79	11.2	24.5	31.4	7.48	392	12.7
80	11.1	24.2	31.0	7.38	387	12.5
81	10.9	23.9	30.6	7.30	383	12.3
82	10.8	23.6	30.2	7.20	378	12.2
83	10.7	23.3	29.9	7.12	373	12.0
84	10.5	23.1	29.5	7.03	369	11.9
85	10.4	22.8	29.2	6.95	365	11.8
86	10.3	22.5	28.8	6.87	360	11.6
87	10.2	22.3	28.5	6.79	356	11.5
88	10.1	22.0	28.2	6.71	352	11.4
89	9.95	21.8	27.9	6.63	348	11.2
90	9.84	21.5	27.6	6.56	344	11.1
91	9.73	21.3	27.3	6.49	341	11.0
92	9.63	21.1	27.0	6.42	337	10.9
93	9.52	20.8	26.7	6.35	333	10.8
94	9.42	20.6	26.4	6.28	330	10.6
95	9.32	20.4	26.1	6.22	326	10.5
96	9.23	20.2	25.8	6.15	323	10.4
97	9.13	20.0	25.6	6.09	320	10.3
98	9.04	19.8	25.3	6.03	316	10.2
99	8.95	19.6	25.0	5.96	313	10.1
100	8.86	19.4	24.8	5.91	310	10.0

Example (i): 63.5 Cotton count = ? Tex

$$63.5 = \frac{63 + 64}{2} \quad \text{or from tables:} \quad \frac{9.37 + 9.23}{2} = 9.30 \text{ Tex}$$

Example (ii): 105 Tex = ? Worsted count
10.5 Tex = 84.4 (from tables)

$$\text{hence } 105 \text{ Tex} = \frac{84.4}{10} = 8.44 \text{ Worsted count.}$$

NOTE:

The values of counts which are not covered by the table (i.e., below 1 and higher than 100) can be calculated as under:

For lower than 1 count, find from tables the value for a count 10 times the known value and multiply the read-off value by 10 to get the desired result.

For higher values, enter the count column at one-tenth of the known value, read off the equivalent and divide by 10. This is explained in example (ii) above.

Index